FACILITY MANAGEMENT
VOLUME II
A PROFESSIONAL GUIDE

RAMESH UPADHYAY
JITENDRA NARAYAN KUMAR

INDIA · SINGAPORE · MALAYSIA

Notion Press

Old No. 38, New No. 6
McNichols Road, Chetpet
Chennai - 600 031

First Published by Notion Press 2019
Revised Edition 2020
Copyright © Ramesh Upadhyay & Jitendra Narayan Kumar 2019
All Rights Reserved.

ISBN 978-1-64733-642-4

This book has been published with all efforts taken to make the material error-free after the consent of the author. However, the author and the publisher do not assume and hereby disclaim any liability to any party for any loss, damage, or disruption caused by errors or omissions, whether such errors or omissions result from negligence, accident, or any other cause.

While every effort has been made to avoid any mistake or omission, this publication is being sold on the condition and understanding that neither the author nor the publishers or printers would be liable in any manner to any person by reason of any mistake or omission in this publication or for any action taken or omitted to be taken or advice rendered or accepted on the basis of this work. For any defect in printing or binding the publishers will be liable only to replace the defective copy by another copy of this work then available.

"A man who is graduated yesterday & stops learning today,
will become uneducated tomorrow"

– Dr. Abdul Kalam

CONTENTS

INTRODUCTION

Facility Management is a professional discipline, which primarily works as a support function. Facility Management mainly focuses on efficient and effective delivery of support services for the organizations that it serves.

The International Organization for Standardization(ISO) defines facility Management as the "organizational function which integrates people, place and process within the built environment with the purpose of improving the quality of life of people and the productivity of the core business." I as an author can add Technology into the above definition.

Facility Management professionals are required to carry all kind of knowledge such as personal management, financials, engineering, hospitality, catering, communication, travel & fleet management, logistics, health & safety, environment, government regulations, compliance, space management, travel & transport, energy Management, equipment & machineries management and technology (IoT) implementation.

Facility Management in our country is bifurcated in two main streams; i.e. Facility Management and Property Management/Infrastructure Management. In India Facility Management is considered as management of office and Property Management is considered as Management of properties such as commercial or residential complex or manufacturing units.

Roles and Responsibility of Facility/Property Managers

Facilities managers operates across various business functions such as Information Technology Company, a BPO, Commercial Complex, Residential Complex, manufacturing industry, industrial warehousing, oil & gas facilities and camps. The number one priority of a Facility Manager is keep the facility operational with minimum resources, high efficiency, no breakdowns, uninterrupted business and with highest level of safety for personal and business. The facility managers are also required to ensure a safe work environment for personnel. Facility managers have to operate in two levels:

- Strategically-tactically: helping clients, customers and end-users understand the potential impact of their decisions on the provision of space, services, cost and business risk.

- Operationally: ensuring corporate and cost effective environment for the occupants to function.

A Facility Manager accomplishes his/her task by managing:

Environment, health and safety

The responsibility of FM department is to ensure a safe work environment for the personnel's and organization. Failure to do so may lead to unhealthy conditions leading to employees falling sick, injury, loss of business, prosecution and insurance claims. The confidence of customers and investors in the business may also be affected by adverse publicity from safety lapses.

Fire & Safety

Fire is the highest risk factor for any organization and a potential fire is risk to loss of life, property and business. The responsibility of facilities management department is to ensure undertake timely safety audit, have right kind of maintenance in place, inspection and testing for all of the fire safety equipment and systems, keeping records and certificates of compliance.

Security

There are two types of security. The first one is physical security and other one is electronic security. Security of employees and the business is the responsibility of Facility Management team. Maintenance of the electronic security hardware is part of Facility Management and controlling and training of manned guarding is also part of facility management department.

Maintenance, Audits & Inspections

Maintenance, audits and inspection are required to ensure that the facility is operating safely and efficiently, to maximize the life of equipment and reduce the risk of failure. It is important to ensure that operations of facility has minimum impact on environment. Planning & timely maintenance of equipment's is to ensure long life and low energy consumption.

Hygiene & Cleaning

To ensure a safe work environment hygiene and cleaning plays an important role. Cleaning operations are often undertaken out of business hours, but provision may be made during times of occupations for the cleaning of toilets, replenishing consumables, litter picking and reactive response is scheduled as a series of periodic (daily, weekly, monthly) tasks.

Operations

The facilities management team has to ensure day-to-day smooth operations of the facilities. There are some issues, which may require more than just periodic maintenance. Employees are considered as customers and to support any kind of issue faced by operational staff, the facilities management "help desk" manages many of these that employees can be contact either by telephone or email. The response to help desk calls are prioritized, complaints on help desk varies from AC cooling too hot or too cold, lights not working, photocopier jammed, coffee spills, lights not working or vending machine problems.

Help desks can also be used for booking of meeting rooms, car parking spaces, travel tickets and many other services, but this often depends on how the facilities department is organized. Facilities function is normally split into two sections, which are often referred to as "soft" services such as reception, mailroom, cleaning services, transport management, travel and "hard" services, such as the mechanical, fire and electrical services.

Business Continuity Planning

For a business to flourish it is important to ensure uninterrupted business. In view of the same and organizations should have a business continuity plan in place. So that in the event of a fire or major failure the business can recover quickly. In large organizations setup an alternate site is always kept ready for business recovery and may be the staff move to another site that has been set up to model the existing operation. The facilities management department would be one of the key players should it be necessary to move the business to a recovery site.

Space Management

Facility Management team is the owner of facilities (office or property) and they are responsible for space allocation and cost allocation to business unit. In many organizations, office layouts are subject to frequent changes. This process is referred as churn, and the percentage of the staff moved during a year is known as the (churn rate). The facilities management department plans the movement of business functions as and when requested by business function. This can be done manually or by help of computer-aided design. In addition to meeting the needs of the business, compliance with statutory requirements related to office layouts include:

- the minimum amount of space to be provided per staff member
- fire safety arrangements
- lighting levels
- signage
- ventilation
- temperature control
- welfare arrangements such as toilets and drinking water and hygienic place for food consumption.
  Consideration should also to be given for pantry and catering

This book is a sincere effort by the authors in collating the information's and sharing with the young facility management professionals. We have referred a good number of BIS standards and NBC and the same is represented here for easy references. We hope our efforts will help the FM fraternity in upgrading their knowledge and it will help in enhancing their performance. This initiative will also help the corporate with an updated work force with required information radially available.

We are also hopeful that the organizations shall also benefit from our efforts and this will help them in reducing their operational cost with increased efficiency of their FM team.

Ramesh Upadhyay

Words from the FM Fraternity

"Let me take the opportunity to congratulate Ramesh on release of his second book. Ramesh a stalwart professional in the field of Facilities Management has played a Pivotal Role in the building the Facilities fraternity within the Indian Industry. Ramesh recognition within the Industry to provide Facilities Management Training to budding Real Estate, Administration & Security professionals is highly admired. The second book in the Series covers the Topics of Risk Management, Sustainability and Health & Safety to the next level in terms of Explanation of the Concept, Detailing to finer areas on the subject and live examples from his experience, which are of great importance for FM Professionals to understand the Theory & practical aspects of Business Management. Ramesh has provided deeper knowledge on subjects such as a Commercials within FM, delve in the critical aspects of Housekeeping, Sanitation & Cafeteria Management. I would urge all the Facilities Management professionals to read this book and use it as ready reckoner in future career progression. All the best to Ramesh & looking forward to have a few more books in the coming years."

Girish Awachat
Director - Real Estate & Workplace Services
Sungard Availability Services, INDIA

This book very aptly describes the future of a new Era of service integration, smart technology and business productivity. It balances the technical and the practical aspects of facilities management in a unique way. It also highlights that Innovation is important to not only achieve a strong competitive position but also helps in retention.

The Author's rich experience of the sector illustrated through numerous examples and anecdotes, making it relatable and a joy to read.

I would recommend this book to anyone with an eye for detailing and sense of management.

Rajesh Pongot,
Vice President
Head Facilities Management
Shapoorji Pallonji Investment Advisors Pvt. Ltd.

Word of Thanks

My Special thanks to:

Vijaya Adyanthaya, Director Pest O Crush

Sanjeev Mago

Sanjeev Gautham

Vishal Kadam

Late Piyush Roy

For their contribution, guidance and all support to complete this book.

ENVIRONMENT HEALTH & SAFETY

What EHS is: (Environment, Health, and Safety)

Environment- Conserving Air, Water, Soil, Plants, Animals, Wildlife, Our Community (causing no damage)

Health- Preserving Human Health both Chronic and Acute (preventing illness)

Safety- Preserving Human and Community Safety/ Well Being (preventing injury)

Intent: Why an EHS Management System

- Legal requirements

- Cost optimization

- CSR

  - Global warming, acid rain, ozone depletion, depletion of natural resources, rapid environmental degradation, increasing air pollution, water pollution, noise pollution

  - Resource conservation

  - Prevention of pollution

  - Eliminate negative impacts & promote positive impacts on environment

- Employee care – Healthy & safe workplace

- Business requirements – Client specifications

- Brand building via certifications etc.

EHS Management Systems

- ISO14001 EMS ENVIRONMENTAL MANAGEMENT SYSTEMS

  - the actual requirements for an environmental management system. It applies to those environmental aspects which the organization has control and over which it can be expected to have an influence.

- OHSAS 18001 is an Occupation Health and Safety Assessment Series

  - for health and safety management systems. It is intended to help an organizations to control occupational health and safety risks. It was developed in response to widespread demand for a recognized standard against which to be certified and assessed

- Clause 4.2 Environmental Policy/ OH&S Policy

  - Appropriate to the nature & scale (OH&S risk, environment Impacts of products & services)

  - Commitment to continual improvement (prevention of pollution)

  - Policy should be available to public

- WHAT DOES OHSAS REQUIRE? :

  – Accidents & nonconformities are reported and corrective actions taken Part of planning must include hazard identification, risk assessment and risk control.

  – Management review Management commitment Continual improvement

EHS Management Systems

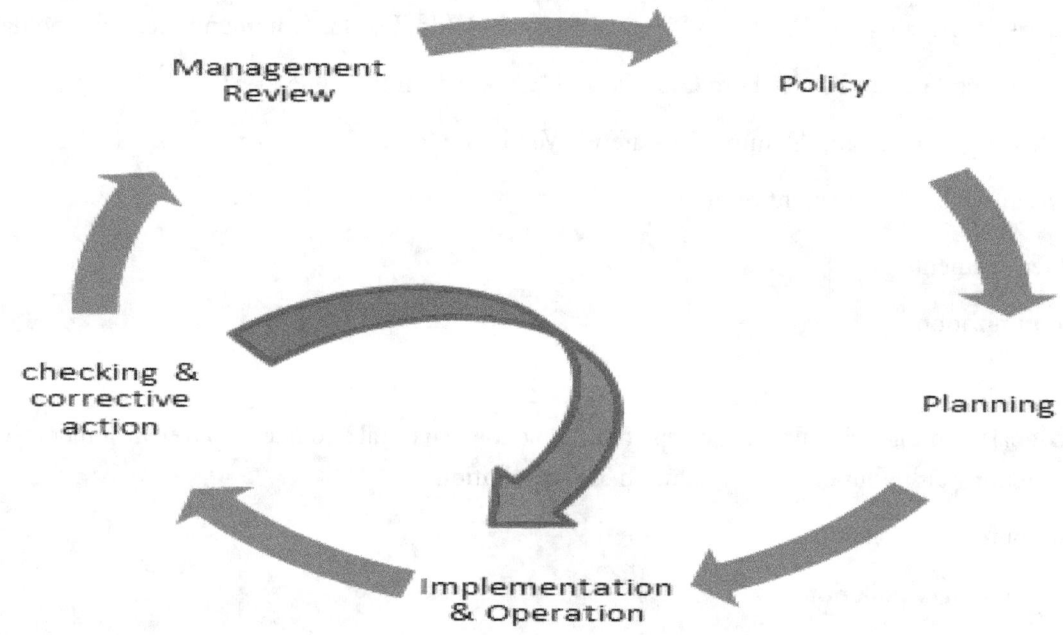

EHS Management System- Benefits

♦ Provides tools to ensure health & safety is everyone's responsibility Integrate health and safety into all aspects of your business

♦ Effective safety and health programs earn positive returns on their health and safety investment by reducing work-related accidents and ill-health and the costs associated with them.

♦ Improving performance through heightened employee morale and adherence to policies and procedures Reinforcing a responsible and well-managed reputation with customers, stakeholders, and communities

♦ Provides a structure to incorporate health and safety into the business Includes health and safety impacts as part of the business process and planning activities and achieving the goal of an accident-free workplace

EHS- Standard Requirements

Safety and Health Inspections

♦ Safety and Health Inspections Conduct regular (usually weekly) site inspections

♦ Establish daily work area inspection procedures

♦ Develop and use a checklist

- Provide a reliable system for employees, without fear of reprisal, to notify management about apparent hazardous conditions

- Receive timely and appropriate responses

Additional Worksite Analysis

- Investigate accidents and "near miss" incidents so that their causes and means for prevention are identified.

- Analyze injury and illness trends, so that common cause patterns can be identified and prevented

Hazard Prevention and Control

- Start by determining that a hazard or potential hazard exists

- Where feasible, prevent hazards by effective design of job or job site.

- If the hazard cannot be eliminated, use hazard controls and mitigation procedures.

- Eliminate or control hazards in a timely manner

Controlling the Hazards

- Engineering controls

- Administrative controls

- Personal protective equipment

- Safe work practices communicated via training

- Positive reinforcement

- Correction of unsafe performance

- Enforcement to prevent and control hazards

Hazard Prevention Planning

- Design and maintain the facility and equipment

- Emergency planning Training and drills

- Medical program First aid on site Physician and emergency care nearby

- Safety and Health Training : Safety and Health Training Address the safety and health responsibilities of all personnel Incorporate it into other training and job performance/practice

- Safety and Health Orientation: Employees must understand the hazards they may be exposed to and how to prevent harm to themselves and others from hazard exposure. Orientation training must be given to site and contract workers

Documentation and Record keeping

- Manuals, Procedures, WI & formats

- Policy, Objectives & Targets, Manual, Procedures, WI, Records.

- Accidents, incidents, non-conformities & corrective & preventive action and deviation register

- Procedures for operational controls

- Document Control

    - Easily retrievable and legible

        - Periodically reviewed, revised & approved by authorized levels

        - Current versions are available at all locations

        - Obsolete documents promptly removed

        - identification, storage, protection, retrieval, retention & disposal of records

Monitoring, Measurement and Audits

- Power consumption, Scrap disposal/ Plastic usage

- Accident/ incident events

- Safety and health register

- Emergency preparedness & response

- Evaluation of compliance (ISO)- Monthly evaluation

- Internal Audit

- External/certification Audits

- Management Review

    Some Environmental Impacts From Our Day-to-Day Work

Input	Activity, Product or Service	Aspect	Impact (Potential Consequence)
Water	Toilet & bathroom Use	Waste Water	Air, Water & Soil Pollution
Paper	Printing	Waste Paper	Soil, Water
Oil & Grease	Maintenance	Spillage, Cotton waste	Water and soil poisoning
Diesel fuel	Engine operation, Storage	Exhaust emission, leakage, spillage	Air, soil & water pollution, Fire
Raw Food Material	Food Preparation In Canteen	Waste & Left over food	Air, Water & Soil pollution,

You can contribute: Be Responsible!

As a part of Environment & community, you are responsible to:

- Perform your job in an environmentally safe and sound manner

- Know how your job impacts the environment

- Adhere to standard operating procedures (SOP)

- Know what the potential environmental impacts are for departing from these SOPs

- Know the environmental legal (and other) requirements of your job

QMS/EMS requirements

QMS	EMS
Quality objectives, Procedures & Work Instructions	Environmental Objectives
List of documents & Records	Aspects identification & there impacts
Responsibility & authority matrix	Legal & Other requirements
Client meeting minutes & action plan	Wastage minimization records
Asset register	Decrease in hazardous waste generation
Planning Records	Energy saving
Contract & amendments	Record of Water, paper, Chemical & Fuel usage
Approved vendor list, Supplier evaluation	Chemicals MSDS batch wise
Supplier's product inspection note	Declaration letter for supplier's commitment towards environment
PO/register	Water testing- drinking/ Toilets/Washrooms
Supplier contracts	Sound level/Air emission for DG set
Inventory of stores/Identification & Traceability	Ambient air quality
Equipment list, History card, Maintenance card	Signage's for energy savings
Customer property	Service records for AC's, DG, etc.
Calibration of instruments	Compressed air leakages records
C S/Appreciation/CSS records	Incident reports for Accidents, water wastage, land spillages, etc.
Record of non - conformity/ Remedy's/AP's	WTP/STP records
Analysis of data/Preventive actions taken	Environmental communications/Programs organized at client premises.

Health Hazards

The different types of health hazards that could be associated to IT related work industry could be

- Visual Display Terminal (VDT) hazards,

- Carpel Tunnel Syndrome (CTS),

- And of course the most prominent, is the Back pain.

Other typical examples would be

- Repetitive strain (RSI)
- Neck strain,
- Back pain,
- Sore joints,
- Postural problems

Hazards at Work Place

- Fire in workplace
- Electrical Shock due to lose wire fittings
- Getting trapped in lifts
- Cut injuries due to sharp edges
- Accidents due to rash driving within the campus
- Dehydration
- Slip and fall due to wet floor

    Cost of an accident

HIDDEN

- Equipment damaged loss
- Material loss
- Production loss

    Time lost from work by injured.

DIRECT

- Lost time accident
- Disablement
- Death

Let's Discuss EHS in detail

Protecting people, communities and the environment are fundamentally important to the way any corporate operates. Every corporate/ organization who is involved in any kind of business has responsibility towards people's health, safety and environment. The organization has responsibility for sustainable operations without affecting the environment.

All the organization should have an EHS policy defined for them and this EHS policy should be promptly displayed at their organization.

EHS Policy should broadly cover following key operating principles:

1. Maintain a safe and healthy working environment for all employees, contractors and guests

2. Foster a culture of EHS excellence that is built on integrity, accountability, collaboration and active employee participation, and seek to continuously improve our systems, processes and standards to further support that culture

3. Investigate and implement approaches to reduce the resources we use during the design, development and manufacturing of our products and delivery of commercial services, so as to minimize our impact on the environment

4. Understand the potential hazards associated with our products and take action to reduce any potential risks or adverse impacts

5. Promote EHS excellence in our supply chain and expect third parties doing work on our behalf to do the same. We enter into business relationships with partners that share our commitment to responsible EHS stewardship

6. The EHS Management System should be based on the "Plan, Do, Check, Act" model, which allows one to assess and continually improve their practices over time:

7. The planning process includes development of goals, objectives and metrics based on a review of company performance, EHS programs, applicable regulations and other external factors [PLAN]

8. Activities are informed by Standards, Guidelines and Tools, which are integrated into the EHS Management System, and include specific expectations for sites and operating organizations [DO]

9. Governance committees, from the executive-level EHS Council through site-compliance committees, review performance and progress against objectives. Central audits and self-assessments raise issues, and our monthly and annual performance metrics reflect our progress [CHECK]

10. Corrective actions and continuous-improvement initiatives are established to resolve EHS concerns that have surfaced during performance reviews, assessments, audits and routine surveillance of the regulatory landscape [ACT]

Training

Training is critical to building worldwide employee competencies that will improve compliance, reduce risks and drive continuous improvement.

The training programs are to be reviewed periodically to ensure they remain current. The EHS training program materials should be available in both instructor-led as well as e-learning formats. A mandatory course for all senior company leaders highlights the importance of EHS to the business, the critical role senior leaders play in EHS performance, and the specific actions the leaders can take to drive their areas of accountability toward EHS excellence.

Organization Responsibility

It is the responsibility of an organization to ensure that their operations doesn't cause injury or death and doesn't affect health of their employees and the residents of neighbors. The organization is also responsible to ensure that because of their operations the environment is not affected and pollution is kept under control. The organization should take utmost care to ensure all the pollutants are scientifically treated and recycled.

HSE management has two general objectives

i.   Prevention of incidents or accidents that might result from abnormal operating conditions and reduction of adverse effects that result from normal operating conditions.

ii. Regulatory requirements such as Factory Act and Air & Water pollution control act play an important role and HSE/Facility managers must identify and understand relevant regulations, the implications of which must be communicated to executive management so the company can implement suitable measures by drafting their own policies and procedures. And avoid any legal action in case of non-compliance.

The concern organization has to have an organized efforts and procedures for identifying workplace hazards and reducing accidents and exposure to harmful situations and substances. It also includes training of personnel in accident prevention, accident response, emergency preparedness, and use of protective clothing and equipment. The organization should also have a systematic approach to comply with environmental regulations, such as managing waste or air emissions to reduce the company's carbon footprint.

Environmental

Human impact on the environment or anthropogenic impact on the environment includes changes to biophysical environments and ecosystems, biodiversity, and natural resources. The impact on environment by humans directly or indirectly includes global warming, environmental degradation (such as ocean acidification), mass extinction and biodiversity loss, ecological crises, and ecological collapse.

Further modification of the environment to fit the needs of society is causing bad effects, which become worse as the problem of human overpopulation continues. Some human activities that cause damage (either directly or indirectly) to the environment on a global scale include human reproduction, overconsumption, overexploitation, pollution, and deforestation, to name but a few. Some of the problems, including global warming and biodiversity loss pose an existential risk to the human race, and overpopulation causes those problems

Human beings affect the environment in some or other way. Weather it is a simple housing complex or a manufacturing or process industry or it is an IT industry, every one contributes in polluting the environment in some or other way. It is responsibility of every individual to ensure that the natural resources are recycled and reused, waste generation is minimized and the pollutants are always under control and within prescribed limits of CPCB norms. As per the India Factory act and Air & Water Pollution Control act one has to ensure following points:

Air Emissions: The smoke generated from the power plant or from the processing units should comply with the CPCB norms. A half yearly stack monitoring is undertaken and finding of any pollutants exceeding limits needs to be immediately actioned.

Emissions from new diesel engines used in generator sets have been regulated by the Ministry of Environment and Forests, Government of India. The regulations impose type approval testing, production conformity testing and labeling requirements. Government has also released list of vendors who are authorized to undertake emission tests on DG sets.

Emission limits for new diesel engines ≤ 800 kW used in generator applications were set in 2002 [3003][3002] and strengthened in 2013 [3001]. The regulations also set noise limits for diesel generator sets up to 1000 kVA.

The emission standards listed in Table 1 summarizes the standards that became applicable in 2004/2005, and Table 2 lists the strengthened limits applicable from April 2014.

Table 1						
Emission standards for diesel engines ≤ 800 kW for generator sets (2004/2005)						
Engine Power (P)	Date	CO	HC	NOx	PM	Smoke
		g/kWh				1/m
P ≤ 19 kW	2004.01	5	1.3	9.2	0.6	0.7
	2005.07	3.5	1.3	9.2	0.3	0.7
19 kW < P ≤ 50 kW	2004.01	5	1.3	9.2	0.5	0.7
	2004.07	3.5	1.3	9.2	0.3	0.7
50 kW < P ≤ 176 kW	2004.01	3.5	1.3	9.2	0.3	0.7
176 kW < P ≤ 800 kW	2004.11	3.5	1.3	9.2	0.3	0.7

Table 2				
Emission standards for diesel engines ≤ 800 kW for generator sets (2014)				
Engine Power (P)	Date	CO	NOx + HC	PM	Smoke
		g/kWh			1/m
P ≤ 19 kW	2014	3.5	7.5	0.3	0.7
19 kW < P ≤ 75 kW	2014	3.5	4.7	0.3	0.7
75 kW < P ≤ 800 kW	2014	3.5	4	0.2	0.7

Engines are tested over the 5-mode as per ISO 8178 D2 test cycle. Smoke opacity must not exceed the limits at full load (Table 1) or at any load point of the test cycle (Table 2). The same limits are applicable for type approval and for conformity of production (COP) testing.

Emission standards for new diesel engines above 800 kW used in generator set applications were phased-in between 2003 and 2005, Table 3 [3004].

Table 3				
Emission limits for diesel engines > 800 kW for generator sets phased-in between 2003 and 2005, Table 3 [3004].				
Date	CO	NMHC	NOx	PM
	mg/Nm3	mg C/Nm3	ppm(v)	mg/Nm3
Until 2003.06	150	150	1100	75b
2003.07 - 2005.06	150	100	970 (710a)	75c
2005.07	150	100	710 (360a)	75c

a) For engines in plants of total power rating above 75/150 MW located in urban/rural areas, respectively.

b) 150 mg/Nm3 for engines fueled with furnace oil.

c) 100 mg/Nm3 for engines fueled with furnace oil.

Concentrations are corrected to dry exhaust conditions at 15% residual O2.

Ambient Air Quality

Air Pollutant includes any substance in solid, liquid or gaseous form present in the atmosphere in concentrations which may tend to be injurious to all living creatures, property and environment. Various contaminants in different forms from both man-made and natural sources perpetually enter our environment and contaminate it causing toxicity, diseases and environmental decay.

Recent industrialization and increased number of potent air polluting sources like automobiles, landfills, etc. has gushed into the atmosphere toxic materials which not only harm human health but are also a threat to the ecosystem in general.

In order to arrest the deteriorating air quality, Government of India enacted Environment (Protection) Act, 1986, which was an umbrella act for the protection of all aspects of the environment. The government had also enacted the Air (Prevention and Control of Pollution) Act, 1981 with the aim of ensuring moderated pollution concentrations and hence safe ambient air quality.

Air pollution is a global concern. One cannot merely abstain themselves from pollution at an individual level because unlike water whose purity can be ensured before consumption, air has to be taken up in the form that it is present. Air pollution is known to cause a lot of premature deaths, chronic asthma, decreased fertility, deteriorated property and may lead to the hazardous acid rains too. It is hence imperative that the air pollutant concentrations are regularly monitored and a proper inventory is maintained so that one can predict unforeseen disasters like The Great Smog of 1952 in London. With the aim of proper monitoring of the pollutant concentrations, the Central Pollution Control Board (CPCB) in1984-1985 launched the National Ambient Air Quality Monitoring network which was later renamed as National Air Monitoring Program (N.A.M.P.).

The objectives of N.A.M.P. can be enlisted as follows-

1. To determine the trend and status of the ambient air quality.

2. To ascertain that the prescribed levels are being met or not

3. To identify the cities where the pollution concentration is higher than the prescribed levels.

4. To understand and obtain knowledge for the development of preventive and corrective measures.

5. To understand the ongoing natural processes of cleansing like dispersion, dilution, dry deposition, wind-based movements, chemical transformation and precipitation.

The objective of National Ambient Air Quality Standards (NAAQS) is-

1. To indicate necessary air quality levels and appropriate margins required to ensure the protection of vegetation, health and property.

2. To provide assistance in the establishment of priorities for abatement and control of pollution.

3. To provide a uniform yardstick for assessment of air quality at the national level.

4. To indicate the extent and need of monitoring program.

National ambient air quality standard: The Clean Air Act requires EPA to set national ambient air quality standards (NAAQS) for carbon monoxide and five other pollutants considered harmful to public health and the environment (the other pollutants are ozone, particulate matter, nitrogen oxides, sulfur dioxide and lead).

An air quality index (AQI) is a number used by government agencies to communicate to the public about how polluted is the air currently or how polluted it is forecast to become. As the AQI increases, an increasingly large percentage of the population is likely to experience increasingly severe adverse health effects.

Ambient air quality parameters

Pollutant	Time weighted average concentration in ambient air				Method of measurement
		Industrial area	Residential, Rural & Other area	Sensitive area	
Sulphur Dioxide (So2)	Annual average	80µg/m3	60µg/m3	15µg/m3	1) Improved west & Gaeke method
	24 hrs.	120µg/m3	80µg/m3	30µg/m3	2) Ultraviolet fluorescence
Nitrogen di-Oxide	Annual average	80µg/m3	60µg/m3	15µg/m3	1) Jacob & Hochheiser method (NaArsenite)
	24 hrs.	120µg/m3	80µg/m3	30µg/m3	2) Gas phase (Chemiluminescence)

Pollutant	Time weighted average concentration in ambient air				Method of measurement
		Industrial area	Residential, Rural & Other area	Sensitive area	
Suspended Particulate method (SPM)	Annual average	360μg/m3	140μg/m3	70μg/m3	
	24 hrs.	500μg/m3	200μg/m3	100μg/m3	Average flow rate not less than 1.1m3/minute
Repairable Particulate matter (size less than 10μm) (RPM)	Annual average	120μg/m3	60μg/m3	50μg/m3	
	24 hrs.	150μg/m3	100μg/m3	75μg/m3	

Pollutant	Time weighted average concentration in ambient air				Method of measurement
		Industrial area	Residential, Rural & Other area	Sensitive area	
Carbon Monoxide (CO)	8 hrs.	5.0 mg/m3	2.0 mg/m3	1.0 mg/m3	Non dispersive infrared spectroscopy
	1 hrs.	10 mg/m3	4.0 mg/m3	2.0mg/m3	Average flow rate not less than 1.1m3/minute
Note 01	Annual arithmetic mean of minimum 104 measurements in a year taken twice a week 24 hrly at uniform interval.				
Note 02	24 hrly/8hrly values should be met 98% of the time in a year. However, 2% of the time, it may exceed but not on two consecutive days.				

The air quality is categorized under 4 broad categories based on Exceedance Factor.

$$\text{Exceedance Factor} = \frac{\text{Observed annual mean concentration of criteria pollutant}}{\text{Annual standard for the respective pollutant and area class}}$$

The four air quality categories are

Critical pollution (C): when EF is > 1.5;

High pollution (H): when the EF is between 1.0 − < 0.5;

Moderate pollution (M) : when the EF between 0.5 − <1.0;

Low pollution (L): when the EF is < 0.5.

| Pollution level | Annual Mean Concentration Range (µg/m3) | | | | | |
| | Industrial, Residential, Rural & others areas | | | Ecologically Sensitive Area | | |
	SO2	NO2	PM10	SO2	NO2	PM10
Low (L)	0-25	0-20	0-30	0-10	0-15	0-30
Moderate (M)	26-50	21-40	31-60	11-20	16-30	31-60
High (H)	51-75	41-60	61-90	21-30	31-45	61-90
Critical (C)	>75	>60	>90	>30	>45	>90

Wastewater and Ambient Water Quality

Wastewater includes process wastewater, wastewater from utility operations, storm water and sanitary wastewater. Wastewater will vary in quality and quantity by industry sector and typically includes:

Process wastewater: Water that comes in contact with any raw material, product, by-product, or waste during any production or industrial **process**.

Pollutants may include acids, bases, and many others. These include soluble organic chemicals, suspended solids, nutrients (phosphorus and nitrogen), heavy metals (such as cadmium, chromium, copper, lead, mercury, nickel and zinc), cyanide, toxic organic chemicals, oily materials and volatile materials. The costs of treating process wastewater can be significant.

Wastewater from utilities operations: Utility operations such as cooling towers and demineralization systems may result in high rates of water consumption, as well as the potential release of high temperature water containing high dissolved solids, residues of biocides and residues of other cooling system anti-fouling agents.

Storm water: Storm water includes any surface runoff and flows from process and materials staging areas resulting from precipitation or drainage. Typically storm water runoff contains suspended sediments, metals, petroleum hydrocarbons, Polycyclic Aromatic Hydrocarbons (PAHs) and coliform. Rapid runoff, even of uncontaminated storm water, also degrades the quality of the receiving water by eroding stream beds and river banks.

Sanitary wastewater: This may include effluents from domestic sewage, food service and laundry facilities serving site employees and can also include other sources such as from laboratories, medical infirmaries, equipment maintenance shops and water softening.

As an responsible corporate house and HSE team it is important to regularly monitor the quality, quantity, sources and discharge points of liquid effluents by type (process, utilities operations, storm water and sanitary). At a facility level, discharges of wastewater should not result in contaminant concentrations in excess of the effluent discharge quality standards. We should works as an responsible citizen of the earth and ensure the effluent is minimized by scientific processing and recycling. There are various means to filter and recycle these water wastes.

Water Treatment Process

a) Phase separation

In phase separation, grease and oil is recovered from the waste water. The oil/grease recovered from separation process can be used as fuel or saponification (Saponification is a process where the waste oil or fats are used for making of soaps).

b)   Sedimentation

Under sedimentation the waste water is collected in a tank which is also known as settling tank. Solid and non-polar liquids are removed from wastewater by gravity. It is also widely used for the treatment of other wastewaters. Solids that are heavier than water will accumulate at the bottom of tank. This tank will also have skimmers to simultaneously remove floating grease like soap scum and solids like feathers or wood chips.

c)   Filtration

The fine solids which can't be removed by sedimentation process, in this case the water is passed through a filter. Other types of water filters remove impurities by chemical or biological processes described below.

Biochemical Oxidation

Oxidation reduces the biochemical oxygen demand of wastewater. Oxidation is also known as aeration of waste water. Bacteria and protozoa consume biodegradable soluble organic contaminants (e.g. sugars, fats, and organic short-chain carbon molecules from human waste, food waste, soaps and detergent) while reproducing to form cells of biological solids. Biological oxidation processes are sensitive to temperature and, between 0 °C and 40 °C, the rate of biological reactions increase with temperature. Oxidation helps in growth of bacteria and protozoa.

Chemical Oxidation

In the Chemical oxidation process some persistent organic pollutants and concentrations remaining after biochemical oxidation are removed. In the process of Chemical oxidation bacteria and microbial pathogens are killed by adding ozone, chlorine or hypochlorite to waste water.

Softening

The treated water contains high level of hardness may be above 200 PPM, which makes this treated water unusable for gardening and construction activity. In view of the same the water needs to be softened by an activated carbon filter followed by sand filter. The process reduces hardness to 100 to 150 PPM.

Water Quality

Water quality testing is an important part of environmental monitoring. When water quality is poor, it affects not only aquatic life but the surrounding ecosystem as well.

These sections detail all of the parameters that affect the quality of water in the environment. These properties can be physical, chemical or biological factors.

Physical properties of water quality include temperature and turbidity.

Chemical characteristics involve parameters such as pH and dissolved oxygen.

Biological indicators of water quality include algae and phytoplankton.

These parameters are relevant not only to surface water studies of the ocean, lakes and rivers, but to groundwater and industrial processes as well.

Discharge Water parameter post treatment as per CPCB Norms (Central Pollution Control Board)

Sr. No.	Parameters	Standards			
		Inland surface water	Public Sewers	Land for Irrigation	Marine area
1	Suspended solids mg/l	100	600	200	For process water 100, for cooling water effluent 10% above total suspended
2	Particulate size of suspended solids	shall pass 850 micron IS Sieve			a) Floatable solids, max. 3mm b) Settleable solids, max 850 microns
3	pH Value	5.5 to 9.0	5.5 to 9.0	5.5 to 9.0	5.5 to 9.0
4	Temperature	Shall not exceed 5 deg C above receiving water temp.			Shall not exceed 5 deg C above receiving water temp.
5	Oil & grease mg/ltrs	10	20	10	20
6	Total residual chloringe mg/ltr	1	-	-	1
7	Ammonical nitrogen (N) mg/ltr	50	50	-	50
8	Total Kjaldahl Nitrogen (NH3) mg/ltr	100	-	-	100
9	Free Amonia (NII3) mg/ltr	5	-	-	5
10	Biochemical Oxygen Demand (3day at 27deg.C) mg/litre	30	350	100	100
11	Chemical Oxygen Demand mg/ltr	250	-	-	250
12	Arsenic (As) mg/ltr	0.2	0.2	0.2	0.2
13	Mercury (Hg) mg/ltr	0.01	0.01	-	0.01
14	Lead (Pb) mg/ltr	0.1	1	-	2
15	Cadmium (Cd) mg/ltr	2	1	-	2
16	Hexvalent Cromium (Cr+6) mg/ltr	0.1	1	-	1
17	Total Chromium (Cr) mg/ltr	2	2	-	2
18	Copper (Cu) mg/ltr	3	3	-	3
19	Zinc (Zn) mg/ltr	5	15	-	15
20	Selenium (Se) mg/ltr	0.05	0.05	-	0.05
21	Nickel (Ni) mg/ltr	3	3	-	5
22	Cynide (Cn) mg/ltr	0.2	2	0.2	0.2
23	Fluoride (Fl) mg/ltr	2	15	-	15
24	Dissolved Phospate (P) mg/ltr	5	-	-	-
25	Sulphide (S) mg/ltr	2	-	-	5
26	Phenoile Compound (C6H5OH) mg/ltr	1	5	-	5
27	Manganese (Mn) mg/ltr	2	2	-	2

(Contd.)

Sr. No.	Parameters	Standards			
		Inland surface water	Public Sewers	Land for Irrigation	Marine area
28	Iron (Fe) mg/ltr	3	3	-	3
29	Vanadium (V) mg/ltr	0.2	0.2	-	0.2
30	Nitrate Nitrogen mg/ltr	10	-	-	20

Waste Management

Waste is a major issue across the globe. Improper waste disposal causes contamination of water bodies, ground water and soil; it also causes harm to the environment.

A waste can be in any form either solid, liquid, or gas. Each waste have different methods of disposal and management. Waste management normally deals with all types of waste whether it was created in forms that are industrial, biological, household, and special cases where it may pose a threat to human health. Waste management is intended to reduce adverse effects of waste on the health of all living beings and the environment.

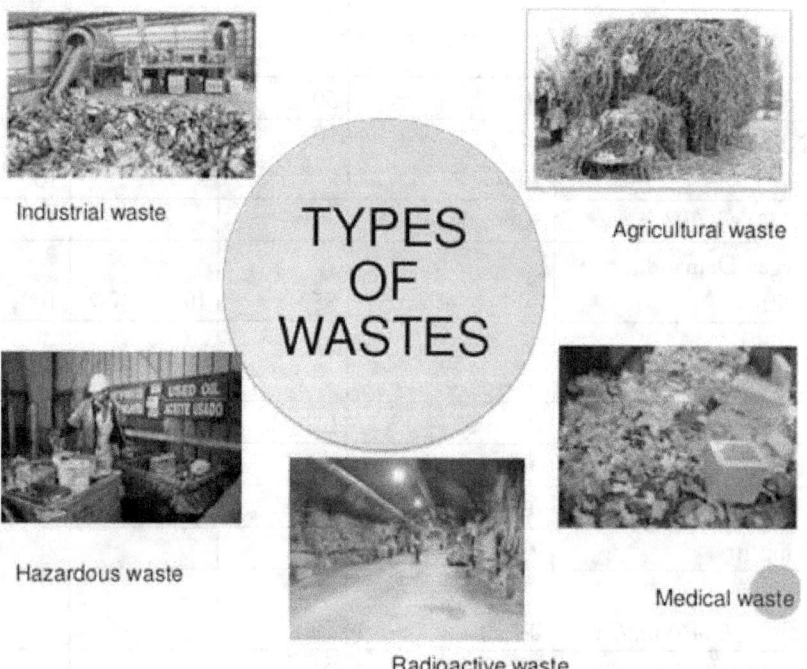

Industrial waste

Agricultural waste

TYPES OF WASTES

Hazardous waste

Medical waste

Radioactive waste

Types of Waste

1. Liquid Waste: Liquid waste is commonly found both in households as well as in industries. This waste includes dirty water, organic liquids, wash water, waste detergents and even rainwater.

2. Solid Rubbish: Solid rubbish can include a variety of items found in your household along with commercial and industrial locations.

    Solid rubbish is commonly broken down into the following types:

    Plastic waste – This consists of bags, containers, jars, bottles and many other products that can be found in your household. Plastic is not biodegradable, but many types of plastic can be recycled. Plastic should not be mix in with your regular waste, it should be sorted and placed in your recycling bin.

Paper/card waste – This includes packaging materials, newspapers, cardboards and other products. Paper can easily be recycled and reused so make sure to place them in your recycling bin or take them to your closest Brisbane recycling depot.

Tins and metals – This can be found in various forms throughout your home. Most metals can be recycled. Consider taking these items to a scrap yard or your closest Brisbane recycling depot to dispose of this waste type properly.

Ceramics and glass – These items can easily be recycled. Look for special glass recycling bins and bottle banks to dispose them correctly.

3. Organic Waste: Organic waste is another common household. All food waste, garden waste, manure and rotten meat are classified as organic waste. Over time, organic waste is turned into manure by microorganisms. However, this does not mean that you can dispose them anywhere.

4. Recyclable Rubbish: Recyclable rubbish includes all waste items that can be converted into products that can be used again. Solid items such as paper, metals, furniture and organic waste can all be recycled. If you're unsure whether an item is recyclable or not, look at the packaging or the diagrams on the lid of your yellow recycling bin. Most products will explicitly state whether they are recyclable or not.

5. Hazardous Waste: Hazardous waste includes all types of rubbish that are flammable, toxic, corrosive and reactive. These items can harm you as well as the environment and must be disposed of correctly. Therefore, I recommend you make use of a waste removal company for proper disposal of all hazardous waste.

Hazardous waste includes electronic waste, Chemical waste, Bio-Medical waste etc.

Waste Disposal: There are many means of waste disposal which is as appended below:

1. **Landfill:** A landfill is a specially designed facility for the burial of municipal solid waste. They are designed in such a way that leachates cannot leak down through the soil into the water table. This is accomplished by a layer of clay-like soil at the bottom of the landfill. The next layer up is a synthetic lining, usually made of plastic. All remaining layers alternate between soil and refuse. The solid waste which is considered non harmful for the environment can be used for landfill, however the municipal corporations (government authorities) are found to be disposing almost all the city garbage in landfills. Almost 70% of total waste generated is sent for landfill.

   We as a responsible citizen of this earth can reduce the waste being sent to landfill by a simple step of reducing, reusing, recycling, and composting waste materials.

   Advantages and Disadvantages of Landfill

Advantages:

Something to do with our waste that cannot be recycled.

Methane gas is produced in a landfill by anaerobic decomposition. It can be collected using current technology and then used to generate electricity, or it can be purified and used as a power-generating fuel.

When landfill sites are closed and become "capped" you often get a lot of wildlife grow on top - many of them become wildlife reserves.

Disadvantages:

Finite space - if you put all waste into landfill, you would soon be struggling to find places to put it all.

Wasteful when not required - if recyclable materials are put into landfill sites you use more energy and resources trying to create new materials instead of re-using or recycling so landfill sites should not be the first port of call for re-usable or recyclable materials.

Landfill if not managed scientifically, will lead to contamination of ground water.

Generated foul smell in the vicinity.

Can cause various kind of disease among local habitants.

Generates foul smell in the surrounding areas.

2. Recycling is a resource recovery practice that refers to the collection and reuse of waste materials such as empty beverage containers, plastic material, metal waste, paper waste. The materials from which the items are made can be reprocessed into new products.

    Biological Reprocessing (De-composting or vermicomposting)

    Kitchen waste, plant material, food scraps, and paper products, can be converted into vermiculture by composting and digestion processes to decompose the organic matter. The resulting organic material is then recycled as mulch or compost for agricultural or landscaping purposes. In addition, waste gas from the process (such as methane) can be captured and used for generating electricity and heat. The intention of biological processing in waste management is to control and accelerate the natural process of decomposition of organic matter.

3. Incineration

    Under the process of incineration the organic substances contained in waste materials are burnt through a scientific method. Incineration of waste materials converts the waste into ash, flue gas and heat. The ash is mostly formed by the inorganic constituents of the waste and may take the form of solid lumps or particulates carried by the flue gas. The flue gases must be cleaned of gaseous and particulate pollutants before they are dispersed into the atmosphere. In some cases, the heat generated by incineration can be used to generate electric power.

    Incineration with energy recovery is one of several waste-to-energy technologies. Incineration and gasification may also be implemented without energy and materials recovery.

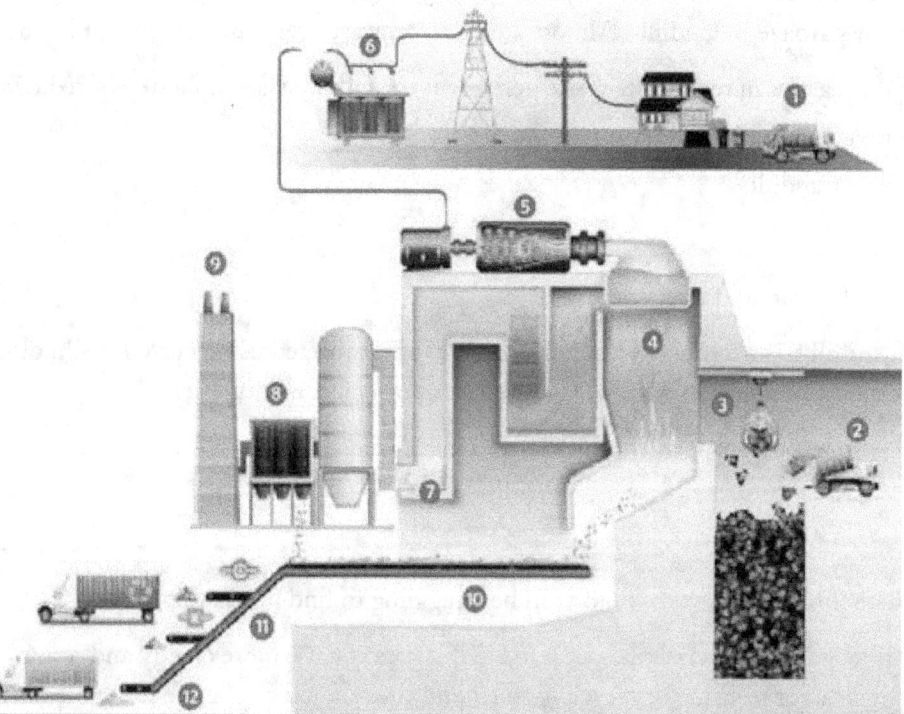

1. Garbage collection

2. Waste delivered at processing plant

3. Waste fed in combustion chamber

4. Heat generated boils water in boiler

5. Steam generated operates power generator

6. Power distribution

7. Gas passed through air pollution control equipment

8. Gas moved through fabric filter to collect particles

9. Emission is monitored through controlled area

10. Ash is collected

11. Metals are collected for recycling

12. Ash reused or disposed of as landfill

Occupational Health and Safety

The motto of Occupational health and safety is to ensure;

i.   Promotion and maintenance of the highest degree of physical, mental and social well-being of workers.

ii.   Prevention of work related health issues caused by their working conditions.

iii.   Protection of workers in their employment from risks resulting from the kind of work they do.

iv.   Placing and maintenance of workers in an occupational environment adapted to physical and mental needs.

In other words, occupational health and safety encompasses the social, mental and physical well-being of workers.

Successful occupational health and safety practice requires collaboration and participation of both employers and workers in a proficiently designed health and safety programs. Under this program issues related to occupational medicine, industrial hygiene, toxicology, education, engineering safety, ergonomics, psychology etc is to be covered.

Most of the organization don't take occupational health issues seriously and have either no or very limited policies in place. The reason behind negligence towards EHS is because the former are generally more difficult to confront and is also a costly affair. It is the responsibility and duty of an organization to ensure that the employees are provided a neat, clean and healthy environment to work. The issues of both health and safety must be addressed in every workplace.

There are many engineering firm who take Environment Health & Safety very seriously. They have all the required policies, training and standards in place. These companies also ensure that their employees adhere to these standards and policies.

Effect of Poor work environment on workers' health and safety

Unhealthy/unsafe working conditions can be found in offices, factories, agricultural fields, mines, construction works, road constructions, service activity and can pose many health and safety hazards.

Occupational hazards can have harmful effects on workers, their families, and other people in the community, as well as on the physical environment around the workplace. A classic example is the use of pesticides in agricultural work or office space for pest control. Workers can be exposed to toxic chemicals in a number of ways when spraying pesticides: they can inhale the chemicals during and after spraying, the chemicals can be absorbed through the skin, and the workers can ingest the chemicals if they eat, drink, or smoke without first washing their hands, or if drinking water has become contaminated with the chemicals. The workers' families can also be exposed in a number of ways: they can inhale the pesticides which may linger in the air, they can drink contaminated water, or they can be exposed to residues which may be on the worker's clothes. Other people in the community can all be exposed in the same ways as well. When the chemicals get absorbed into the soil or leach into groundwater supplies, the adverse effects on the natural environment can be permanent.

Overall, efforts in occupational health and safety must aim to prevent industrial accidents and diseases, and at the same time recognize the connection between worker health and safety, the workplace, and the environment outside the workplace.

Let's consider a case of HVAC (Air conditioning) in office area. If the AHU/Indoor air filters are not maintained properly, this will lead to ingress of dust which is also known as Particulate Matter (PM5 & PM10). These dusts are harmful for human health. Also if the carpet in office is not regularly cleaned and vacuumed may lead to health issue for many employees.

Importance of occupational health and safety

Work plays a central role in people's lives, since most workers spend at least eight hours a day in the workplace, whether it is on a plantation, in an office, factory, etc. Therefore, work environments should be safe and healthy. Yet this is not the case for many workers. Every day workers all over the world are faced with a multitude of health hazards, such as:

- dusts;
- gases;
- noise;
- vibration;
- insufficient PPE
- extreme temperatures.
- Physical work activity involving various tools and equipment's.

Due to noncompliance of HSE standards, lack of attention and hazards involved in works, work-related accidents and diseases are common in all parts of the world.

An employer should understand that it is a moral and legal responsibility of them for the protection of workers' health and safety.

Costs of occupational injury/disease

Work-related accidents or diseases are a costly affair and it directly affect the individual, his/her family and the organization. Cost of occupational accidents has many serious direct and indirect effects on the lives of workers, their families and the organization.

For workers some of the direct costs of an injury or illness are:

- the pain and suffering of the injury or loss of limbs;
- the loss of income;
- the possible loss of a job;
- health-care costs.

Direct impact of an accident for an Organization:

- Loss of trained resource
- Bad name with customer and in market
- Loss of time and delay in work schedule
- payment for work not performed;
- medical and compensation payments;
- repair or replacement of damaged machinery and equipment;
- reduction or a temporary halt in production;
- increased training expenses and administration costs;
- possible reduction in the quality of work;

- negative effect on morale in other workers.
- Legal case

Some of the indirect costs for employers are:

- the injured/ill worker has to be replaced;
- a new worker has to be trained and given time to adjust;
- it takes time before the new worker is producing at the rate of the original worker;
- time must be devoted to obligatory investigations, to the writing of reports and filling out of forms;
- accidents often arouse the concern of fellow workers and influence labor relations in a negative way;
- poor health and safety conditions in the workplace can also result in poor public relations.

Health and Safety Program

It is important that the employers, workers and unions are committed to health and safety and ensure that:

- The workplace hazards are controlled - **at the source** by undertaking JSA (Job Safety Analysis);
- Records of hazard exposure are maintained and a suitable mitigation plan is ready in place.
- Workers and employers should be aware of health and safety risks in the workplace;
- An organization should have an active and effective health and safety committee that includes both workers and management;
- Workers health and safety efforts are ongoing.

Effective workplace health and safety programs can help to save the lives of workers by reducing hazards and their consequences. Health and safety programs also have positive effects on both worker morale and productivity, which are important benefits. At the same time, effective program can save employers a great deal of time, resources and revenue.

Waste Collection and Color Coding

Waste collection and handling is an important part, the employee handling the waste is always in risk of being affected by infection or injection or injection due to contaminate or hazardous waste.

The waste collector should use complete PPE when handling any kind of waste and as an facility manager its our responsibility to ensure safe working environment for our housekeeping staff or the waste handlers. A PPE will be rubber gloves or leather gloves depending on kind of waste being handled, safety shoes and M 3 facemask.

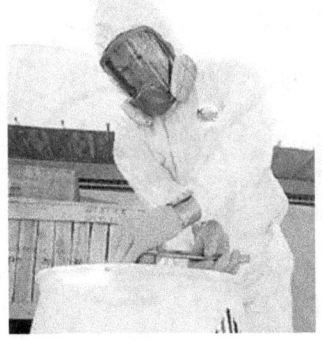

Person handling hazardous waste

Person Handling General waste with PPE

It is important to find out what kind of waste is to be handled and accordingly the person should put on PPE. Hazardous and infected waste is to be handled by trained personal only.

By law, it is mandatory to ensure the waste, is segregated at its source and disposed. The hazardous waste is to be handed over to only authorized vendors by Central or State Pollution control board. In return one should ensure that you receive form 10 for your record.

Waste disposal color-coding

To sure that you always dispose of waste correctly, all bins, bags and containers for waste disposal are color-coded.

Yellow is for clinical waste.

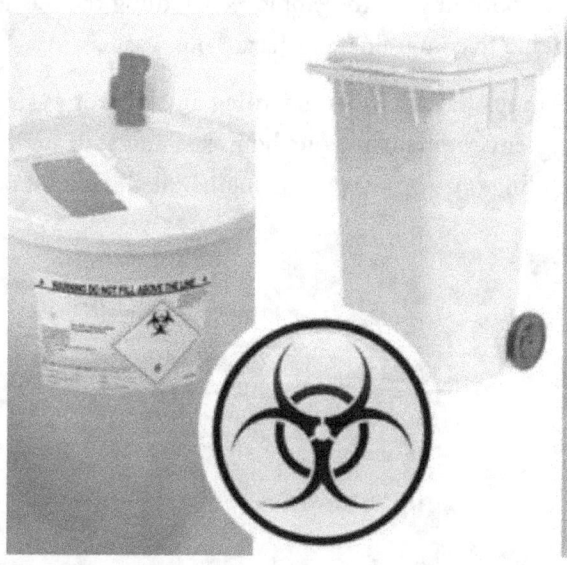

Infected Waste

Used PPE (Personal protective equipments)   **Dispose in yellow bag**

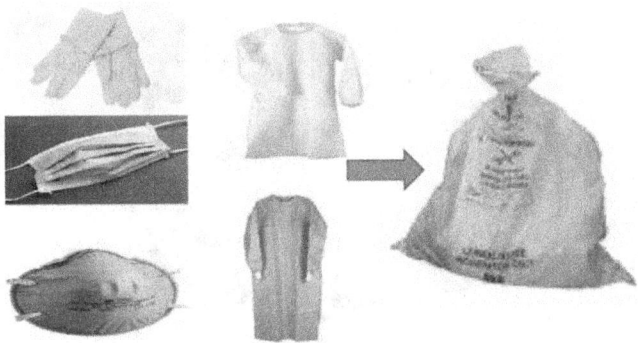

PPE for Infected waste

Clinical waste containers are also marked with the biohazard symbol, just to remind everyone that their contents are hazardous.

Green/Black/Buff/White is for general waste.

General waste is generally not hazardous, and does not contain contaminated items. Standard disposal procedures are sufficient for dealing with general waste.

Blue is for recyclable waste

Blue bin are used for recyclable waste such as metal or plastic.

Purple is for cytotoxic waste.

Cytotoxic waste containers are also marked with the cytotoxic symbol. This type of waste is very hazardous and must be handled by people who had specialist training.

Red is for radioactive waste.

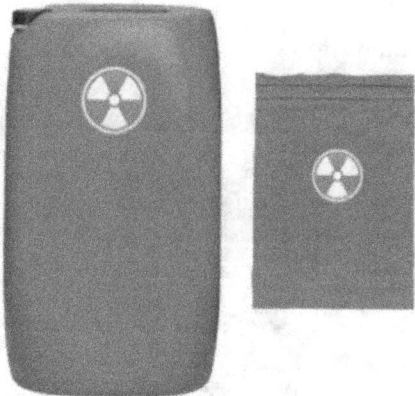

Radioactive waste containers are also marked with the radiation symbol. Radioactive waste is very hazardous and special disposal procedures will be in place. Only people who are qualified and properly trained can dispose of radioactive waste.

Note: Hazardous chemicals – which include radiographic, pharmaceutical, chemical and amalgam wastes – must be disposed of as per state/territory government requirements

DOCUMENTATION

Other important process under Health and Safety Program the organization should follow. The processes are as appended below:

1. Facility Risk Register

A risk register is a document used as risk management tool and to fulfill regulatory compliance acting as a repository for all risks identified and includes additional information about each risk, e.g. nature of the risk, reference and owner, mitigation measures. A Risk Register plots the impact of a given risk over of its probability.

ISO 73:2009 Risk management defines a risk register to be a "record of information about identified risks".

It is important for an organization to ensure a "Facility Risk Register" is prepared and maintained all the time. The Risk Register is required to be updated every 02 years.

Definitions

Hazard - Any real or potential condition that can cause injury, illness, or death to personnel; damage to or loss of a system, equipment or property, or damage to the environment.

Job Safety Analysis (JSA) - A safety management tool in which the risks or hazards of a specific job in the workplace are identified, and then measures to eliminate or control those hazards are determined and implemented.

Probability - The determination of how likely an event or exposure will actually occur in the future.

Risk - Undesirable situation or circumstance that has both a likelihood of occurring and a potentially negative consequence including injury, loss of revenue, assets, market share or damage to the company image.

Risk assessment - Process of evaluating the risk(s) arising from a hazard(s), taking into account the adequacy of any existing controls, and deciding whether or not the risk(s) is acceptable.

Severity - The degree of impact of the impact for an actual or potential event.

Risk assessments should be conducted for the following

A) Prior to commencing a new job or service.

B) Introduction of a new process or significant equipment.

C) In response to a High Potential incident.

D) In response to a significant incident where a new hazard is determined to not have sufficient or effective controls.

E) In response to changes in regulations that affects the worksites.

A site specific Risk/Hazard Assessment program shall be developed and include the following steps and/or components:

A) Identify team; Establish the risk assessment team and responsible Supervisor/Designee, and Manager/Service Coordinator.

B) Identify risks; Identify applicable workplace or task specific hazards that need to be planned for, controlled, or mitigated.

C) Evaluate risks; Evaluate the currently existing risks (severity/ likelihood and initial risk priority code (RPC) utilizing the Risk Analysis Matrix. Refer to the Global HSE Guideline, Hazard Identification and Risk Assessment.

D) Develop risk controls; Define the most effective risk control measures to mitigate the currently existing risks and reduce the RPC to an acceptable level.

E) Access residual risk; Reassess the risks with the new risk control measures in place.

F) Review and approve risk controls; Supervisor/Management or designee review and approval.

G) Implement risk controls.

H) Communicate risks and controls; Communication implementation to the employees performing the work and/ or exposed to the applicable hazards.

I) Evaluate; Monitor the effectiveness of the implemented risk controls.

"Facility Risk" in the "Category" field, has 8 associated generic Facility Risks (and "Other") which are available to select:

A) Fire

B) Electricity

C) Potable Water

D) Food Services

E) Facility Maintenance

F) Statutory Authorizations & Permitting

G) Facility Security

H) Natural Hazards

Task/Activity: For the Real Estate Facility Risk Register, the identified tasks for each generic risk are:

Fire – Operate Facility

Electricity – Operate Facility

Potable Water – Provide Water for Drinking & Personal Hygiene

Food Services – Provide Food Services for Company Employees

Facility Maintenance – Operate Facility

Construction & Statutory Authority Permitting – Perform Construction and or Facility Improvements

Asbestos – Operate Facility

Facility Security – Operate Facility

Natural Hazards – Operate Facility

Other – insert associated task/activity as applicable

Health & Safety Risk Matrix

This will lead to a copy of HSE and SQ Risk Matrices for determining the Risk Priority Code (RPC) using the severity and probability of each possible cause and consequence, and also offer guidance on the possible targets of each risk.

What is impacted (Target)

This section identifies what would be impacted, or the target associated with each risk. That is, if the risk were to occur, what would be affected?

Targets include:

P - Personnel,

E - Equipment,

DT- Downtime,

DC - Data,

Env - Environmental,

I – Interface

Initial Risk – Severity/Probability/RPC

Using the risk matrix, identify the Severity if the situation occurred (catastrophic, critical, marginal, negligible) and the likelihood or probability of that situation happening (frequent, reasonably probable, occasional, remote, extremely improbable, remote).

These together (the severity and probability) will generate an RPC of 1, 2 or 3.

RPCs (Risk Priority Code) are classified as follows:

1 – High Risk: Imperative to suppress risk to lower level

2 – Medium Risk: Operation may require waiver endorsed by management

3 – Operation Permissible

Note that a RPC of less than 3 is NOT ACCEPTABLE for hazards that target personnel.

Risk Reduction Control Measure to Reduce Risk

These can been pre-populated by the management team with required risk reduction control measures based on companies facility management process & procedures and/or requirements of HSE standards, Company Policies & Business Practices. Please add others if required according to local circumstances.

Residual Risk

Once the risk reduction control measures are implemented the residual risk is assessed, what impact does this have on the RPC? (Bear in mind that the Severity cannot be changed. If the severity is classed as catastrophic if the event occurred, that would be true no matter what. Only the likelihood of that event occurring can have an impact on the RPC code for incidents with catastrophic consequences). Insert RPC for residual risk once all the associated risk reduction control measures are complete.

Contingency Plan

This section is optional and free-form and shall be used to document any items where local contingency plans may be required. Note that Contingency planning is a follow on to risk assessment. Contingency plans are based on assessed risks identified during the risk assessment. Contingency planning is not required for every risk. It is required when identified risk cannot be prevented or mitigated to an acceptable level. Contingency plans shall be identified within the risk assessment, actions documented & plans developed locally as required.

Hazard Checklist

Hazard Checklist						
Chemical	Drop Objects	Use of equipment's	Ignition source	Vibration	Mechanical Lifting	Lighting
Drowning	Slips, Trips, Fall	Hand tools	Explosion	Radiation	Operation of vehicle	Weather
Biological agent	Manual Handling	Pressure	Hot/Cold surface	Excavations	Electricity	Electronic waste
Confined Space	Incorrect Posture	Stored Energy	Noise	Height	Environment	Hazardous Chemical

Facility Risk

Targets					
P=Personnel	E=Equipment	DT=Downtime	DC=Data	Env=Environment	I=Interface
Severity					
1-Catastrophic	2. Critical	3-Marginal	4-Negligible		
Probability Rating			**Risk priority Code (RPC)**		
Level	Description		Code	Action Required	
A	Frequent; likely to occur several time		1	High Risk- Mandatory to suppress to a lower level	
B	Reasonable probability; Likely to occur several times		2	Medium Risk- Operation may require waiver by Management to continue	
C	Occasionally; Likely to occur some times		3	Operation Permissible	
D	Remote: Possible but not likely				
E	Extremely improbable; occurrence can't be distinguished Zero				

Risk Matrix

Hazard Severity Category	Description	Service Delivery	A Frequent	B Reasonably Probable	C Occasional	D Remote	E Extremely Possible	F Impossible
I	Catastrophic	1. Potential Equipment failure resulting in significant CPI, COPQ 2. Limited competent personnel available to perform functions/ services that require advanced technical skills. 3. Lack of equipment or facility 4. Single source supplier cannot supply critical spares within the required time frame		1				
II	Critical	1. Higher down time 2. Potential quality escape 3. Lack of competent personal for support 4. Limited supplier to provide critical component			2			
III	Marginal	1. Potential increase in PPM 2. Potential delivery impact resulting in change in mode of transportation. 3. Few suppliers available for critical components				3		
IV	Negligible	1. No impact on quality system or performance 2. No impact on deliver						

Format for Risk Register

Risk Assessment Team Members	Lead and Member 1 are mandatory for the Risk Assessment Team		Tota RA Count :	1 1
Lead *	Member 1 *	Member 2	Member 3	
Member 4	Member 5	Member 6	Member 7	

 # 1 - Risk Assessment

Responsible		Category : Facility Risk		Sub Category : Natural Hazards	
Legal Requiremen	The relevant QHSE Standards include (but not limited to) Hazard Identification & Risk Assessment Standard XXXX; Risk Assessment Standard XXXXX. Where local/country regulations or customer requirements are more stringent than the relevant Global HSE standards, the local/country regulations or customer requirements shall supersede the Halliburton Global HSE standards. Where local/country regulations or customer requirements provide additional requirements, the local/country regulations or customer requirements shall be supplemented to the relevent Halliburton Global HSE standards [insert local/country/customer specific requirements]				

Task/Activity	Risks associated with Task/Activity	How could it happen? (Cause/Aspect)	What are the consequences? (Effect/impact)	What is Impacted (Target)	Initial Risk SEV	Initial Risk PROB	Initial Risk RPC	RISK REDUCTION CONTROL MEASURE TO REDUCE RISK	Residual Risk* SEV	Residual Risk* PROB	Residual Risk* RPC	CONTINGENCY PLAN*
Operate Facility	Naturally occurring events such as severe windstorms, freeze earthquake, flood etc. that could have a negative effect on people and/or the facility	Failure to anticipate and prepare for such events and associated facility risks	Loss of physical Assets; Business Risk; Equipment Damage; HSE Risks; Uncontrolled Radioactive Exposure; Facility Fire; Dropped Objects	[ ] P [ ] E [ ] DT [ ] DC [ ]Env [ ] I [v] P; E; DT; DC; Env	1	C	1	New Properties- Evaluate Natural Hazards (NatHaz) Risks During Property Discovery phase; Undertake NatHaz Identify Associated Actions Assign Actions; Track Close Actions. Develop,Document & Agree Input for Local Emergency Response Plan (LERP) with Local Stakeholders and Communicate LERP Requirements to Impacted Parties as Appropriate. Annually or Following Significant Undertake NatHaz Risk Review. Associated CRE document to Manage Location & Facility Risk - Natural Hazards;	3	D	3	Evacuation plan and Emergency Response Team is listed out with there required actions to be carried out at the time, we conduct Emergency drills at the site to brief and tackle the situations during eventuality to all employees, New employees are given HSE Induction and briefed with ERP and Muster point

 # 2 - Risk Assessment

Responsible		Category : Facility Risk		Sub Category : Facility Maintenance	
Legal Requirements	The relevant QHSE Standards include (but not limited to) Work Environment; Overhead, Jib, Gantry & Monorail Cranes. Where local/country regulations or customer requirements are more stringent than the relevant HSE standards, the local/country regulations or customer requirements shall supersede the HSE standards. Where local/country regulations or customer requirements provide additional requirements, the local/country regulations or customer requirements shall be supplemented to the relevent HSE standards [insert local/country specific requirements in relation to inspections & maintenance of buildings plant & equipment]				

Task/Activity	Risks associated with Task/Activity	How could it happen? (Cause/Aspect)	What are the consequences? (Effect/impact)	What is Impacted (Target)	Initial Risk SEV	Initial Risk PROB	Initial Risk RPC	RISK REDUCTION CONTROL MEASURE TO REDUCE RISK	Residual Risk* SEV	Residual Risk* PROB	Residual Risk* RPC	CONTINGENCY PLAN*
Operate Facility	Critical Plant & Equipment Failures; Accidental Property Damage & General Wear & Tear; Non-compliance with Applicable Govermental Regulatory Requirements; Lifting Equipment Failure;	Failure to Identify and/or Comply with Local/Country Regulations; Failure to Adequately Maintain Facility and/or Identify Business & HSE Critical Plant & Equipment For Inclusion Within Planned Preventative Maintenance Program; Failure to Complete Corrective Maintenance and or Snow/Ice Clearing	Personal Injuries; Business Interruption; Increased Services Quality & HSE Risks; Penalties & Fines Dropped Objects Lifting Equipment Failure	[ ] P [ ] E [ ] DT [ ] DC [ ]Env [ ] I [v] E; DT; DC; Env; P	2	B	1	Governmental/ Statutory Authority Information; Identify Business, HSE & Security Critical Plant & Equipment. Develop & Implement PMP That Prevents Property Damage, Assures Personnel Safety & Security & Compliance With Regulatory & HSE lifting & Hoisting Equipment- Inspections, Develop & Communicate Facility & Corrective Maintenance Requests Procedure. Associated CRE Facility Preventative Maintenance; Facility & Corrective Maintenance Requests	3	C	3	Maintenance check list is made for all the equipment's in the facility and PPM routines are carried out as per schedule which covers Electrical systems, Machineries, plumbing,civil etc are undertaken on timely basis to monitor timely repairs

 # 3 - Risk Assessment

| Responsible | | Category : Facility Risk | | Sub Category : Fire | | | | | | | | | |

Legal Requirements: The relevant HSE Standards include (but not limited to) Facility Fire Prevention; Fire Equipment Services; Where local/country regulations or customer requirements are more stringent than the relevant Global HSE standards, the local/country regulations or customer requirements shall supersede the HSE standards. Where local/country regulations or customer requirements provide additional requirements, the local/country regulations or customer requirements shall be supplemented to the relevent HSE standards

Task/Activity	Risks associated with Task/Activity	How could it happen? (Cause/Aspect)	What are the consequences? (Effect/impact)	What is Impacted (Target)	SEV	PROB	RPC	RISK REDUCTION CONTROL MEASURE TO REDUCE RISK	SEV	PROB	RPC	CONTINGENCY PLAN*
Operate Facility	Facility Fire	Electrical Faults, Hot Work, Naked Flame; Flammable & Combustible Materials; Reactive Chemicals; Cooking & Preparing Hot Food	Evacuation of the Facilities; Personal Injury; Property Damage; Business Interruption	[ ] P [ ] E [ ] DT [ ] DC [ ]Env [ ] I [v] E; DT; DC; Env; P	1	C	1	Undertake Fire Risk Analysis; Identify Associated Actions; Track & Close Actions. Associated CRE Documents: Facility Fire Risk Analysis Process; Facility Fire Risk Analysis Process; Fire Risk Analysis Job Aid; Facility Fire Detection & Suppression Systems	3	D	3	ERP is in place, Timely Evacuation drill is conducted, during this evaluation ERT team take responsibility and performs actions required at time of emergency, fire extinguish training is been conducted quarterly by HSE and CRE

 # 4 - Risk Assessment

| Responsible | | Category : Facility Risk | | Sub Category : Electricity | | | | | | | | | |

Legal Requirements: The relevant HSE Standards include (but not limited to) Electrical Safety; Where local/country regulations or customer requirements are more stringent than the relevant HSE standards, the local/country regulations or customer requirements shall supersede the HSE standards. Where local/country regulations or customer requirements provide additional requirements, the local/country regulations or customer requirements shall be supplemented to the relevent Halliburton Global HSE standards [insert local/country specific requirements]

Task/Activity	Risks associated with Task/Activity	How could it happen? (Cause/Aspect)	What are the consequences? (Effect/impact)	What is Impacted (Target)	SEV	PROB	RPC	RISK REDUCTION CONTROL MEASURE TO REDUCE RISK	SEV	PROB	RPC	CONTINGENCY PLAN*
Operate Facility	Facility Electrical System/Equipment Faults	Overcurrent; Exposed Live Parts; Contact with Overhead/Underground Cables; Over-loading; Damaged Cables; Hazardous Environments	Facility Fire; Electric Shock; Arc Flash Hazard; Equipment Damage; Business Interruption	[ ] P [ ] E [ ] DT [ ] DC [ ]Env [ ] I [v] E; DT; DC; P	1	C	1	Undertake Electrical Risk Analysis; Identify Associated Actions; Assign Actions; Track & Close Actions. Associated CRE Documents: Facility Electrical Inspection Process; Facility Electrical Inspection Guideline; Facility	3	D	3	Electrical Infrastructure maintenance is undertaken along with audit of the system to find out any issues with the system. Thermography is conducted half yearly to safeguard the man material and property with all types of

 # 5 - Risk Assessment

| Responsible | | Category : Facility Risk | | Sub Category Potable Water | | | | | | | | | |

Legal Requirement: The relevant HSE Standards include (but not limited to)Disease Protection, Potable Water, Where local/country regulations or customer requirements are more stringent than the relevant HSE standards, the local/country regulations or customer requirements shall supersede the HSE standards. Where local/country regulations or customer requirements provide additional requirements, the local/country regulations or customer requirements shall be supplemented to the relevent HSE standards [insert local/country/customer specific requirements]

Task/Activity	Risks associated with Task/Activity	How could it happen? (Cause/Aspect)	What are the consequences? (Effect/impact)	What is Impacted (Target)	SEV	PROB	RPC	RISK REDUCTION CONTROL MEASURE TO REDUCE RISK	SEV	PROB	RPC	CONTINGENCY PLAN*
Provide Water for Drinking & personal hygiene	Contaminated Potable Water Systems	Unreliable Muncipal Supply; Local Wells or Surface Sources; Contaminants including Chemicals, Viruses & Bacteria; Cross Contamination; Water Temperatures within range 25-42 c or 77 - 108 f; Design of Plumbing System (infraquent Use; dead-end lines, eye-wash stations etc.); Plumbing Materials (shower heads, washers, hoses etc.); Cooling Towers; Artictectural Fountains & Waterfalls; Water Storage Tanks etc.; Drinks Dispensing	Legionella; Water Borne illnesses and Microbiological Disease;	[ ] P [ ] E [ ] DT [ ] DC [ ] Env [ ] I [v] Env; P	2	C	1	Undertake Risk Analysis Potable Water & Legionells; Deveop Control Plan Including Leves of Monitoring, Frequency of Maintenance; and any Associated Procedures Assign Actions; Track & Close Actions. Associated CRE Documents:[Legionella Guidance Document IP]	3	D	3	Drinking water is being sourced from a branded Mineral water company with test reports which are timely documented.

6 - Risk Assessment

Responsible		Category : Facility Risk				Sub Category Food Services						

Legal Requirement	The relevant CRE HSE Standards include (but not limited to)Food Services; Potable Water; where local/country regulations or customer requirements are more stringent than the relevant Global HSE standards, the local/country regulations or customer requirements shall supersede the Halliburton Global HSE standards. Where local/country regulations or customer requirements provide additional requirements, the local/country regulations or customer requirements shall be supplemented to the relevent Halliburton Global HSE standards [insert local/country/customer specific requirements]

Task/Activity	Risks associated with Task/Activity	How could it happen? (Cause/Aspect)	What are the consequences? (Effect/impact)	What is Impacted (Target)	SEV	PROB	RPC	RISK REDUCTION CONTROL MEASURE TO REDUCE RISK	SEV	PROB	RPC	CONTINGENCY PLAN*
					Initial Risk				Residual Risk*			
Provide Food Services for Company Employees	Contamination or Spoilage of Food During Storage and Preparation	Poor food Hygiene Practices; Inappropriate Storage Conditions and Temperatures; Poor Pest Control; Adequate Supply of Potable Hot Water and Soap for Washing Kitchen Utensils and Hygiene; Poor Health Reporting & Surveillance; Poor Waste Management Procedures; Expired Shelf Life	Illness; Food Poisoning	[v] P [ ] E [ ] DT [ ] DC [ ] Env [ ] I	2	C	1		3	D	3	Food services is not being provided by the company, The Staff house Kitchen is being audited quarterly by HSE and CRE recommendations are given for timely improvement

7 - Risk Assessment

Responsible		Category : Facility Risk				Sub Category Facility Maintenance						

Legal Requirements	The relevant CRE HSE Standards include (but not limited to)Work Environment, Preventative Maintenance, Lifting & Hoisting Equipment - Inspections. Where local/country regulations requirements are more stringent than the relevant CRE HSE standards, the local/country regulations requirements shall supersede the CRE HSE standards. Where local/country regulations requirements provide additional requirements, the local/country regulations requirements shall be supplemented to the relevent CRE HSE standards [insert local/country/customer specific requirements in relation to inspections & maintenance of building plant & equipment]

Task/Activity	Risks associated with Task/Activity	How could it happen? (Cause/Aspect)	What are the consequences? (Effect/impact)	What is Impacted (Target)	SEV	PROB	RPC	RISK REDUCTION CONTROL MEASURE TO REDUCE RISK	SEV	PROB	RPC	CONTINGENCY PLAN*
					Initial Risk				Residual Risk*			
Operate Facility	Critical Plant & Equipment Failures; Snow & Ice Clearance; Severe Wind Storms; Accidental property Damage & General Wear & Tear; Non-compliance with Applicable Governmental Regulatory Requirements; Lifting Equipment Failure;	Failure to Identify and/or Comply with Local/Country Regulations; Failure to Adequately Maintain Facility and/or Identify Business & HSE Critical Plant & Equipment For inclusion Within Planned Preventative Maintenance Program; Failure to Complete Corrective Maintenance of Critical Plant & Equipment For inclusion Within Planned Preventative Maintenance Program; Failure to Complete Corrective Maintenance	Personal Injuries; Business Interruption; increased HSE Risks; Penalties & Fines	[ ] P [ ] E [ ] DT [ ] DC [ ] Env [ ] I [v] E; DT; DC; Env; P	2	C	1	Develop Register of vendor employees who are considered as food handlers; indicate that employees are aware of the requirements of companies food handling policy and any additional governmental food safety regulations. Validate & retain evidence of third party food service vendors awareness and compliance with the requirements of any additional governmental food safety regulations; undertake periodic inspections and ensure any negative findings are recorded & corrected. Associated CRE documents: Vendor Management; Establish Source of Governmental/ Statutory Authority Information;identify Business, HSE & Security Critical Plant & Equipment. Develop & implement PMP That Prevents Property Damage, Assures Personnel safety & security & compliance with regulatory & CRE HSE Lifting & Hoisting Equipment - Inspections Develop & Communicate Facility & Corrective Maintenance Requests Procedure. Associated CRE Documents: Facility Preventative Maintenance; Facility Corrective maintenance Requests	3	D	3	Maintenance Check list is made for all the equipment's, maintenance routines are carried out as per PPM schedule which covers electrical systems, machineries, plumbing, civil etc

8 - Risk Assessment

Responsible		Category : Facility Risk				Sub Category Construction & Improvements Authorizations						

Legal Requirements	The relevant CRE HSE Standards include (but not limited to) Environmental Authorizations/Permitting; HSE record retention; Record control; Where local/country regulation requirements are more stringent than the relevant CRE HSE standards, the local/country regulations or customer requirements shall supersede the companies HSE standards. Where local/country regulations requirements provide additional requirements, the local/country regulations requirements shall be supplemented to the relevent CRE HSE standards

Task/Activity	Risks associated with Task/Activity	How could it happen? (Cause/Aspect)	What are the consequences? (Effect/impact)	What is Impacted (Target)	SEV	PROB	RPC	RISK REDUCTION CONTROL MEASURE TO REDUCE RISK	SEV	PROB	RPC	CONTINGENCY PLAN*
					Initial Risk				Residual Risk*			
Perform construction and or Facility improvements	Non-compliance with applicable governmental regulatory requirements	Failure to Identify and/or renew authorizations required; Late applications for authorizations; failure to comply with conditions of authorizations	increased Business Risk; increased HSE Risks; Penalties & fines; adverse media attention	[ ] P [ ] E [ ] DT [ ] DC [ ] Env [ ] I [v] DT; Env; P	2	C	1	Develop & retain register of construction & facility improvement permits; Regularly review documentation to ensure that all conditions and Limitations of the applicable authorizations are being met as required (at least semi-annually). Ensure construction & facility improvements statutory authorizations & permitting are identified & obtained at the appropriate stage for all new Construction and/or improvement projects. Associated Documents: Construction & facility improvements statutory authorizations & permitting process.	3	D	3	The country CRE team ensures all authorizations are in place befoe starting of any construction or modifications

 # 9 - Risk Assessment

Responsible		Category : Facility Risk		Sub Category Facility Security									
Legal Requirements	The relevant Company Business Practice's include Facility Security. Where local/country regulations are more stringent than the relevant Company Business Practice/ CRE HSE standards, the local/country regulations or customer requirements shall supersede the Halliburton Company Business Practice/CRE HSE standards, where local/country regulations provide additional requirements, the local/country regulations or customer requirements shall be supplemented to the relevent Halliburton Company Business Practice/CRE HSE standards												

Task/Activity	Risks associated with Task/Activity	How could it happen? (Cause/Aspect)	What are the consequences? (Effect/impact)	What is Impacted (Target)	Initial Risk			RISK REDUCTION CONTROL MEASURE TO REDUCE RISK	Residual Risk*			CONTINGENCY PLAN*
					SEV	PROB	RPC		SEV	PROB	RPC	
Operate Facility	Security Threats and Losses Associated with Critical Assets (including Physical, Intellectual & Date)	Failure to Identify & Manage security, threats & risks; Failure to adequately maintain Facility Security	Loss of Physical assets; Business risk; increased personal security	[ ] P [ ] E [ ] DT [ ] DC [ ] Env [ ] I [v] E; DT ;DC; P	2	C	1	New Properties - Evaluate security threat level/Risks during property discovery phase; Undertake & document security analysis for each site/ building; Develop & document site/ building security plan. Get it approved from corporate security; Regularly reveiw facility security plan (at least annually).	3	E	3	The Facility has been rpobvided with electronic and physical security system. The building has a three layer physical secdurity system in place.

Action Plan

	Action Plans				
S. No.	Description of Action	Accountable Person	% Complete	Due Date	Comments /Status
1	Develop Input For Local Emergency Response Plan				[Mark complete when content of LERP template is agreed and incorporated into local LERP]
2	Undertake NatHaz Risk Analysis				[Mark complete when no significant risk of Natural hazards exist or when completed copies of appropriate NatHaz Risk Analysis are uploaded]
3	Undertake Facility Fire Risk Analysis				[Mark complete when complete copy of facility fire risk analysis is uploaded]
4	Identify Fire Risk Control Actions; Assign Actions; Track & Close Actions				Once completed mark complete
5	Undertake Facility Electrical Inspection				Mark complete when completed copy of facility electrical risk analysis ready
6	Identify Electrical Risk Control Actions; Assign Actions; Track & Close Actions				Once completed mark complete
7	Undertake Potable Water & Legionella Risk Analysis; Develop Control Plan Including Levels of Monitoring, Frequency of Maintenance; and Associated Procedures				Mark complete when potable water & legionella risk analysis is ready and approved from Management & control measures identified have been included in PMP calendar

(Contd.)

8	Develop Register of employees Who are Considered Food Handlers, Validate That Employees are Aware of the Requirements of food hygiene and handling and Any Additional Governmental Food Safety Regulations and Have Received Training in Safe Food handling. Identify Refresher training frequencies. Validate & Retain Evidence of Third Party Food Service Vendors Awareness and Compliance With the food safety requirements and Any Additional Governmental Food Safety Regulations				[Mark complete when food services are provided or register of employees and/or vendor personnel shows initial training report is filed with refresher training dates identified]
9	Undertake Periodic Inspections of Food Preparation & Service Areas and Ensure Any Negative Findings Are Recorded & Corrected Food Service Periodic Inspection Checklist For an Example Inspection Checklist)				[Mark complete when food services are not provided or initial inspection has been undertaken and filed with schedule for periodic inspections]
10	Identify Business & HSE Critical Plant & Equipment. Develop & Implement Preventative Maintenance Plan (PMP).				Mark complete when business & HSE critical plant & equipment is identified in PMP and filed.
11	Develop Contingency Plan to Address Short-Term Fire Warning System Outages				[Only required where identified risk cannot be prevented or mitigated to an acceptable level; if required insert specific actions]
12	Develop Contingency Plan to Address Short-Term Power Outages and/or Loss of Power in Emergency Situations				[Only required where identified risk cannot be prevented or mitigated to an acceptable level; if required insert specific actions]
13	Develop & Communicate Facility & Corrective Maintenance Requests Procedure.				[Mark complete when Facility & Corrective Maintenance Requests is implemented and communication plan or copy of communication to employees is filed]
14	Develop & Retain Register of Construction & Facility Improvement Permits				[Mark Complete when register of permits is complete & filed. including schedule for semi-annual review]
15	Undertake Facility Security Risk Analysis				Mark complete when completed copy of facility security risk analysis is ready and filed.
16	Develop Facility Security Plan				Mark complete when facility security plan is developed and copy of approval by Corporate Security Group is filed.

Extent of the problem worldwide

Hazard: The terms "hazard" and "risk" are often used interchangeably. However, in terms of risk assessment, they are two very distinct terms. A hazard is any agent that can cause harm or damage to humans, property, or the environment. Risk is defined as the probability that exposure to a hazard will lead to a negative consequence, or more simply, a hazard poses no risk if there is no exposure to that hazard.

1. There is an unlimited number of hazards that can be found in almost every workplace. These include both obvious unsafe working conditions and insidious, less obvious hazards.

2. Hazards often are built into the workplace. Therefore, trade unions must ensure that hazards are removed, rather than trying to get workers to adapt to unsafe conditions.

3. The most effective accident and disease prevention begins when work processes are still in the design stage, when safe conditions can be built into the work process.

1. Accidents

In general, health and safety in the workplace has improved in most **industrialized** countries over the past 20 to 30 years. However, the situation in developing countries is relatively unclear largely because of inadequate accident and disease recognition, record-keeping and reporting mechanisms.

It is estimated that at least 250 million occupational accidents occur every year worldwide. 335,000 of these accidents are fatal (result in death). (Since many countries do not have accurate record-keeping and reporting mechanisms, it can be assumed that the real figures are much higher than this.) The number of fatal accidents is much higher in developing countries than in industrialized ones. This difference is primarily due to better health and safety program, improved first-aid and medical facilities in the industrialized countries, and to active participation of workers in the decision-making process on health and safety issues. Some of the industries with the highest risk of accidents worldwide are: mining, agriculture, including forestry and logging, and construction.

Identifying the cause of an accident

In some cases, the cause of an industrial injury is easy to identify. However, very often there is a hidden chain of events behind the accident which led up to the injury. For example, accidents are often indirectly caused by negligence on the part of the employer who may not have provided adequate worker training, or a supplier who gave the wrong information about a product, etc. The consistently high fatal accident rates in developing countries emphasize the need for occupational health and safety education program that focus on prevention. It is equally important to promote the development of occupational health services, including the training of doctors to recognize work-related diseases in the early stages.

2. Diseases

Hazard Exposure in the workplace may lead to serious illness

Some occupational diseases have been recognized for many years, and affect workers in different ways depending on the nature of the hazard, the route of exposure, the dose, etc. Some well-known occupational diseases include:

- asbestosis (caused by asbestos, which is common in insulation, automobile brake linings, etc.);

- silicosis (caused by silica, which is common in mining, sandblasting, etc.);

- lead poisoning (caused by lead, which is common in battery plants, paint factories, etc.);

- and noise-induced hearing loss (caused by noise, which is common in many workplaces, including airports, and workplaces where noisy machines, such as presses or drills, etc. are used).

There are also a number of potentially crippling health problems that can be associated with poor working conditions, including:

- heart disease;

- musculoskeletal disorders such as permanent back injuries or muscle disorders;

- allergies;

- reproductive problems;

- stress-related disorders.

Many developing countries report only a small number of workers affected by work-related diseases. These numbers look small for a variety of reasons that include:

- inadequate or non-existent reporting mechanisms;

- a lack of occupational health facilities;

- a lack of health care practitioners who are trained to recognize work-related diseases.

Because of these reasons and others, it is fair to assume that in reality, the numbers of workers afflicted with occupational diseases are much higher. In fact, overall, the number of cases and types of occupational diseases are increasing, not decreasing, in both developing and industrialized countries.

Identifying the cause of occupational disease

The cause of work-related diseases is very often difficult to determine. One factor is the latency period (the fact that it may take years before the disease produces an **obvious** effect on the worker's health). By the time the disease is identified, it may be too late to do anything about it or to find out what hazards the worker was exposed to in the past. Other factors such as changing jobs, or personal behaviors (such as smoking tobacco or drinking alcohol) further increase the difficulty of linking workplace exposures to a disease outcome.

Although more is understood now about some occupational hazards than in the past, every year new chemicals and new technologies are being introduced which present new and often unknown hazards to both workers and the community. These new and unknown hazards present great challenges to workers, employers, educators, and scientists, that is to everyone concerned about workers' health and the effects that hazardous agents have on the environment.

The range of hazards

There are an unlimited number of hazards that can be found in almost any workplace.

The unsafe working conditions are such as:

a) unguarded machinery,

b) slippery floors

c) inadequate fire precautions,

But there are also a number of categories of insidious hazards (that is, those hazards that are dangerous but which may not be obvious) including:

- chemical hazards, arising from liquids, solids, dusts, fumes, vapors and gases;

- physical hazards, such as noise, vibration, unsatisfactory lighting, radiation and extreme temperatures;

- biological hazards, such as bacteria, viruses, infectious waste and infestations;

- psychological hazards resulting from stress and strain;

- hazards associated with the non-application of ergonomic principles, for example badly designed machinery, mechanical devices and tools used by workers, improper seating and workstation design, or poorly designed work practices.

Most workers are faced with a combination of these hazards at work.

Importance of management commitment

Safety & The Supervisor

A successful health & safety program requires strong management commitment and worker participation

In order to develop a successful health and safety program, it is essential that there be strong management commitment and strong worker participation in the effort to create and maintain a safe and healthy workplace. An effective management addresses all work-related hazards, not only those covered by government standards.

All levels of management must make health and safety a priority. They must communicate this by going out into the worksite to talk with workers about their concerns and to observe work procedures and equipment. In each workplace, the lines of responsibility from top to bottom need to be clear, and workers should know who is responsible for different health and safety issues.

1. Strong management commitment and strong worker involvement are necessary elements for a successful workplace health and safety program.

2. An effective management addresses all work-related hazards, not only those covered by government standards, and communicates with workers.

The importance of training

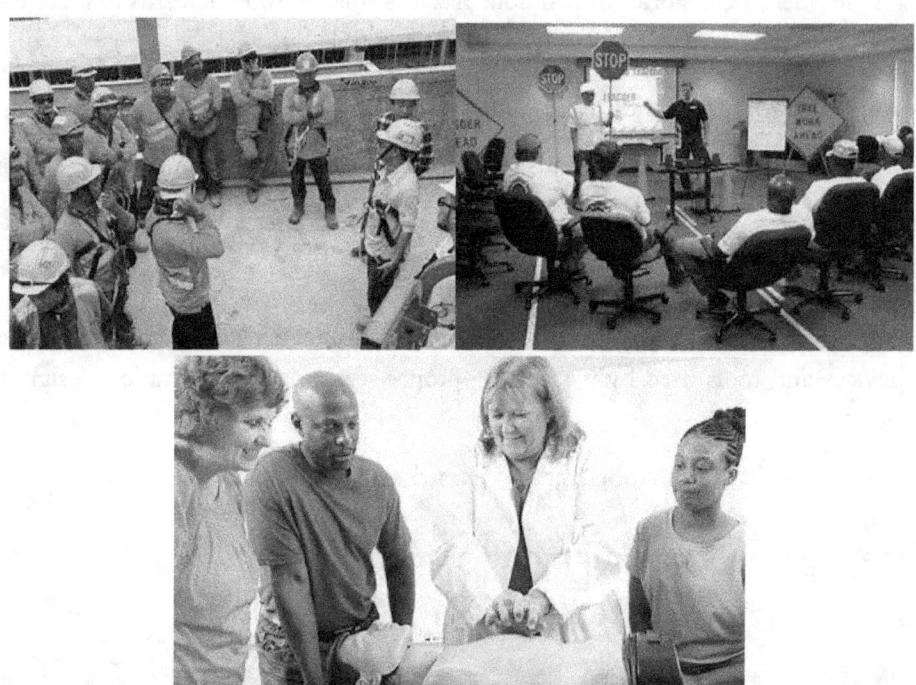

Effective training is a key component of any health & safety program

Workers often experience work-related health problems and do not realize that the problems are related to their work, particularly when an occupational disease is in the early stages. Besides this other more obvious benefits of training are as skills development, hazard identification and mitigation, first aid in case of medical emergencies etc. a comprehensive training program in each workplace will help workers to:

- recognize early signs/symptoms of any potential occupational diseases before they become permanent conditions;

- assess their work environment;

- insist that management make changes before hazardous conditions can develop.

Role of the health and safety officer

Hazard will always be available as part and parcel of a work. Role of health and safety officer is to work proactively (this means taking action **before** hazards become a problem) to prevent workers from being exposed to occupational hazards. You can do this by making sure management eliminates hazards or keeps them under control when they cannot be eliminated.

Being a health and safety officer is not always easy, but helping to protect the lives of your fellow workers is worth all the time and effort you put into the job.

Steps to help you reach your goals are:

1. Be well informed about the various hazards in your workplace and the possible solutions for controlling those hazards.

2. Work together with your vendor and employees to identify and control hazards.

Health & Safety Officer: HSE Officer or Manager should use a variety of source of information about potential or existing hazards in your workplace.

1. **Observe your work place**

2. **Examine Records**

3. **Listen to complaints**

4. **Inspect your workplace**

5. **Ask members what is their view**

6. **Read or collect information**

Identifying hazards in the workplace

1. Welder: A welder can get burn injury from the sparks generated due to welding and also there will be danger of fire due to the work process. The intense light generated due to welding can cause permanent eye injury (may lead to blindness) as well the fumes emanating from the welding works can harm the welder's lungs also.

2. Mechanic: Depending on the precise nature of a mechanic's duties, there may be safety problems from cuts, pinch point, slips and falls, etc., and exposure to chemical hazards such as oils, solvents, asbestos and exhaust fumes. Mechanics can also have back and other musculoskeletal problems from lifting heavy parts or bending for long periods.

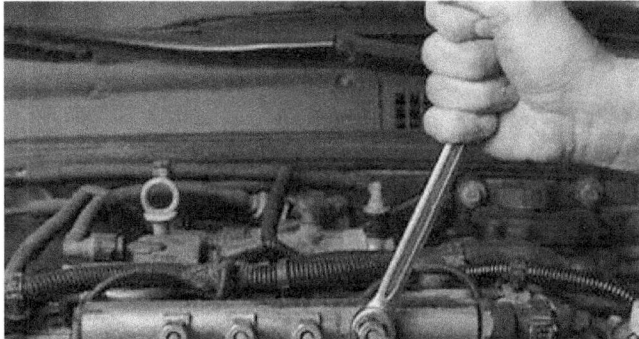

3. Port worker: There are major risks involved with respect to cargo handling at port. The port workers often have no idea of the dangerous nature of the cargo; there may be a sign on the side of a box or drum, but the information may not be in their language or in words that make much sense to the average worker. The condition of the cargo is also important as leaking drums or split bags can be very hazardous for the handlers. Other risks include falls, cuts, back and other musculoskeletal problems as well as collisions with fast moving vehicles such as fork-lift trucks or delivery trucks.

4. Textile worker: Textile worker are prone to many kind of hazards such as threat from machines around them, fire with so much combustible material in the workplace, noise and vibration, dust from the material which can seriously affect the lungs, exposure to cotton dust can lead to the occupational disease known as byssinosis.

5. Agricultural worker: Hazard related to farm workers is with respect to spraying of pesticides on crops, because of this the worker is exposed to hazardous chemicals contained in the spray. Many pesticides and herbicides that have been banned in some countries because of their toxic effects are still used in many developing countries. If spraying takes place on a windy day, the spray can be breathed into the lungs and blown on to the skin where it can cause damage. It can also be absorbed into the body through the skin.

6. Electronics assembly worker: An electronics assembly worker can suffer eye problems from doing close work, often in poor light. Because such workers sit still for long periods with inadequate seating, they can also suffer from back and other musculoskeletal problems. For some workers there are the dangers of solder fumes or solder "flecks" in the eye when the excess solder is cut off with pliers.

7. Office worker: Office workers are prone to health hazard due to stress, ergonomics, exposure to chemical hazards from office machines such as photocopiers. Poor lighting, noise and poorly designed chairs and stools can also present problems.

8. Construction worker: Construction site has plenty of hazards such as falls, slips, trips, cuts, and drop objects (being hit by falling objects). There are also dangers from working at height, working in confined space, musculoskeletal problems from lifting heavy objects, as well as the hazards associated with exposure to noisy machinery.

9. Miner: Miners are prone to many hazards such dusts, fire, explosion and electrocution, as well as the hazards associated with vibration, extreme temperatures, noise, slips, falls, cuts, etc.

Measures to control Hazard

There are various available means and checks, using which you can avoid or eliminate or reduce the hazard. Using of PPE helps in reducing the hazard impact, checks and precautions helps us in avoiding the hazard and limits its impact on work and workers. Implementation of appended work methods and best practices can have an accident free work place.

Works Permit: There are various kind of work permit such as Confined Space permit, working at height, Hot work permit and cold work permit. The permit form is filled to ensure that the concern authorities are aware that what kind of work is undertaken and what all safety precautions are being taken. The work permit needs to be closed after completion of work.

a)   Hot Work/Confined Space work permit form

CONFINED SPACE ENTRY / HOT WORK PERMIT
IN CASE OF EMERGENCY CALL _____

DATE ISSUED: __/__/__ TIME ISSUED: _____ ☐ am ☐ pm   DATE EXPIRES: __/__/__   TIME EXPIRES: _____ ☐ am ☐ pm

Safe Work Permit No: _____ (REQUIRED FOR ALL PERMITS)

WORK LOCATION:	
DESCRIPTION:	

LOCATION	O_2(19.5-23.5%):	LEL (0-10%):	Toxic Measured and ppm	Calibration Current
			_____ _____ ppm	YES ☐
			_____ _____ ppm	YES ☐
			_____ _____ ppm	YES ☐

Safety Attendant ☐   **Fire Watch** ☐

Location: 1st: _____	Emergency Communication Method (e.g.. Air horn/ hand held Radio) YES ☐
Location: 2nd: _____	Colored vest YES ☐
Location: 3rd: _____	Emergency escape routes identified YES ☐
Location: 4th: _____	Attendant analyzer calibrated/sticker current YES ☐

☐ **Confined Space Entry Permit**	☐ **Hot Work Permit**

Confined Space Entry Permit side:

TEMPERATURE INDEX: _____ (0°C / 32°F–26°C / 79°F)

☐ AC / Heater ☐ Work/Rest Regiment Required ☐ Other

When temperatures are outside (0°C / 32°F–26°C/79°F) contact the safety / IH team for further guidance in implementing protective measures.

	YES	N/A
1. Energy Isolation Plan	☐	☐
2. Forced Ventilation Required	☐	☐
3. Low Voltage Equipment	☐	☐
4. GFCI Required	☐	☐
5. Entry / Exit identified	☐	
6. Safety Attendant Log	☐	
7. Bottle Watch Attendant	☐	☐

Hot Work Permit side:

TYPE OF FIRE PROTECTION: QUANTITY
Dry Chemical/CO_2 Extinguisher   _____
Water Hose   _____

	YES	N/A
1. Energy Isolation Plan	☐	☐
2. Ventilation adequate for type of work	☐	
3. Inert purge required / precautions taken	☐	☐
4. GFCI Required	☐	☐
5. Is equipment or piping to be worked on verified?	☐	
6. Fire blankets required	☐	☐
7. Combustibles removed or protected	☐	
8. Flumes/drains/pump seals in area covered	☐	☐
9. Spark Containment Enclosure required	☐	☐
10. Welding machine and ground attachment in safe location	☐	☐

Additional safety considerations or contractor requirements:

I have inspected the job site and verified that all conditions of the applicable permits are met.

ISSUER	RECEIVER/QUALIFIED PERSON

SAFETY ATTENDANTS

DEBRIEFING ON CONFINED SPACE ISSUES (comments):

b) Cold Work Permit:

SAFE WORK PERMIT

Issued (Date/Time): _____   Expire (Date/Time): _____   Unit: _____   Equipment #: _____

| Related Work Permits: | Hot Work/Confined Space ☐ | Electrical ☐ | Excavation ☐ | Other Permit ☐ No: _____ |

DESCRIPTION: _____

_____

Lock Box # (optional): _____   Work Notification # (optional): _____   Vehicle Entry Only ☐

Work Group/Contractor: _____   Job Contact of non-resident contractor: (Name) _____

Safety Concerns		YES	N/A	COMMENTS
1.	Has field verification occurred	☐		Includes job site, energy isolation, & required inspections.
2.	Process equipment cleared, depressured and isolated (EIP) OR Extra precautions listed	☐	☐	Purged With:
3.	Bleeders open and unplugged. Adjacent bleeders located and plugged if necessary	☐	☐	
4.	First break involved?	☐	☐	*Minimum First Break PPE:* Hard Hat, Goggles, Face Shield, Chemical Splash Suit, Chemical Resistant Gloves and Chemical Resistant Boots. Refer to the PPE grid or comments below
5.	LOTO Try been performed?	☐	☐	
6.	Radiation source Locked-Out?	☐	☐	
7.	Equipment grounded to prevent static build up	☐	☐	
8.	Other Work Groups affected (Signatures required below)	☐	☐	(Signatures required below)
9.	Potential for falling objects to a lower level	☐	☐	
10.	Regulated or Hazardous Chemicals Identified?	☐	☐	Specify: ☐Asbestos ☐Lead ☐CrVI Other_____
11.	Critical Instruments Identified?	☐	☐	
12.	Slip, Trip and Fall concerns	☐	☐	
13.	Hot Surfaces/Tracing Locations Identified/Shut Off or protected?	☐	☐	
14.	Required warning barricade/signs installed?	☐	☐	Type: ☐ Danger ☐Caution ☐Other_____
15.	Nearest Safety Shower, Eyewash Tested and Fire Extinguisher identified?	☐		Location:
16.	Emergency Assembly Area Identified?	☐		Primary: Secondary:
17.	Emergency Communication Method Identified	☐		Specify:
18.	Fall Protection Required?	☐	☐	Type:
19.	Respiratory Protection Required? Type?	☐	☐	Specify: ☐Supplied Air ☐Air Purified Respirator_____
20.	Breathing air bottle/trailer inspected?	☐	☐	
21.	Safety Attendant Needed?	☐	☐	Specify Reason:
22.	What Specific Protective Equipment, beyond standard PPE is required for the work. Specify types and when required in comments section.	☐		☐Chemical Resistant Suit ☐Chemical Resistant Boots ☐Face Shield ☐Personal Monitor (eg. CO, H₂S, other) ☐Proper Gloves ☐Hearing Protection ☐Goggles ☐Other_____
23.	Inert gases in the area that require additional precautions?	☐	☐	Precautions:
24.	Chemical Concerns	☐	☐	Specify:
Other concerns or contractor requirements:				

| Detector Reading for Instrument Work in a Classified Area and vehicle entry | LEL: _____ (0-5%) O₂ _____ (19.5-23.5%) Toxics: _____ |

Sign On: We ensure that the precautions checked and/or written above have been taken, and will be followed for the duration of the job.

Work Group/Contractor Representative: _____   Owning Area Representative: _____

Shift Change Re-validation:
Work Group/Contractor Representative: _____   Owning Area Representative: _____
Affected Area Representatives: _____

Job Follow-Up: Signing indicates all work group personnel are released from this Safety Checklist protection.

Jobsite Clean	☐ YES ☐ NO	Safety hazards / comments:
Job Completed	☐ YES ☐ NO	
Work Group/Contractor Representative: _____		Owning Area Representative: _____

Job Safety Analysis

A job safety analysis (JSA) is a procedure which helps integrate accepted safety and health principles and practices into a particular task or job operation. In a JSA, each basic step of the job is to identify potential hazards and to recommend the safest way to do the job. Other terms used to describe this procedure are job hazard analysis (JHA) and job hazard breakdown.

The terms "job" and "task" are commonly used interchangeably to mean a specific work assignment, such as "operating a grinder," "using a pressurized water extinguisher," or "changing a flat tire." JSAs are not suitable for jobs defined too broadly, for example, "overhauling an engine"; or too narrowly, for example, "positioning car jack."

Benefits of doing a Job Safety Analysis

One of the methods used in this example is to observe a worker actually perform the job. The major advantages of this method include that it does not rely on individual memory and that observing or performing the process prompts the recognition of hazards. For infrequently performed or new jobs, observation may not be practical.

Template - JSA/RA

Job Title/Task:					XXXXX					
REGION	Department Xxx	Sub Function xxx			DOCUMENT NUMBER: XXXXX					
OWNER xxx	APPROVED BY: Xxx	REVISION BY: xxx			REV. DATE: XXX	REV. No: D	Final RPC 0			
REQUIRED PPE: Hard Hat, Safety Glasses, Coveralls, Safety Boots, Gloves, Hearing Protection				BEFORE	Remarks: Visually inspect all equipment and tools prior to use.				AFTER	
List of Permits Required: „ Hot Work, „ Cold work, „ Confined space entry, „ Access, „ Electrical, „ Excavation, „ Other			TARGET	SEV	PROB	RPC		SEV	PROB	RPC
Sequence of Steps	Potential Hazard	Hazard Effect					RISK REDUCTION MEASURES			
1										
2										
3										
4										
5										
6										
	Name (print)	Sign	Date:			Job Position:				
	Name (print)	Sign	Date:			Job Position:				
	Name (print)	Sign	Date:			Job Position:				
Review the risk assessment and identify any additional hazard & control measures in this section for the specific task to be carried out. All personnel involved in risk assessment need to print, sign, date & record their job position.										

Note: Review JSA prior to work scope

One approach is to have a group of experienced workers and supervisors complete the analysis through discussion. An advantage of this method is that more people are involved in a wider base of experience and promoting a more ready acceptance of the resulting work procedure. Members of the health and safety committee must also participate in this process.

Initial benefits from developing a JSA will become clear in the preparation stage. The analysis process may identify previously undetected hazards and increase the job knowledge of those participating. Safety and health awareness is raised, communication between workers and supervisors is improved, and acceptance of safe work procedures is promoted.

A JSA, or better still, a written work procedure based on it, can form the basis for regular contact between supervisors and workers. It can serve as a teaching aid for initial job training and as a briefing guide for infrequent jobs. It may be used as a standard for health and safety inspections or observations. In particular, a JSA will assist in completing comprehensive accident investigations.

HIRA (Hazard Identification And Risk Analysis)

Hazard Identification and Risk Analysis (HIRA) is a process used in identifying hazards and evaluating risk at facilities, throughout their life cycle, and to ensure that the risks to personal, the public, or the environment are consistently controlled within the organization's risk tolerance. This is a study undertaken by HSE Officer and the Facilities team to address three main risk questions to a level of detail commensurate with analysis objectives, life cycle stage, available information, and resources.

The three main risk questions are:

➢ Hazard – What can go wrong?

➢ Consequences – How bad could it be?

➢ Likelihood – How often might it happen?

HIRA encompasses the entire spectrum of risk analyses, from qualitative to quantitative. HIRA is important process to manage risk, hazards identification, and then the risks should be evaluated and determined to be tolerable or not. The earlier in the life cycle that effective risk analysis is performed, the more cost effective the future safe operation of the process or activity is likely to be. The risk understanding developed from these studies forms the basis for establishing most of the other process safety management activities undertaken by the facility. An incorrect perception of risk at any point could lead to either inefficient use of limited resources or unknowing acceptance of risks exceeding the true tolerance of the company or the community.

A HIRA study is to be undertaken only by a team of trained and qualified experts on the process, the materials, and the work activities. A simple early-in-life hazard identification study may be performed by a single expert; however, a multidiscipline team typically conducts more hazardous or complex process risk studies, especially during later life cycle stages. Involving operating and maintenance personnel early in the review process will help identify hazards when they can be eliminated or controlled most cost effectively. When the study is complete, management must then decide whether to implement any recommended risk reduction measures to achieve its risk goals.

Risk Assessment Form

Format No.:

| | Location | | Prepared by: | | |
| Process | | | Approved by: | | |

NO	Activity	Sub Activity / Task	Risk / Hazard Identification	Risk Analysis Assess	Risk Control	Identified By	Residual Risk

Sample HIRA Form

Hazard Hunt:

Hazard hunt is an activity where a group of responsible employees undertake joint inspection of facility and machineries. And inspect every aspect of safety in the facility. Hazards are identified and recorded.

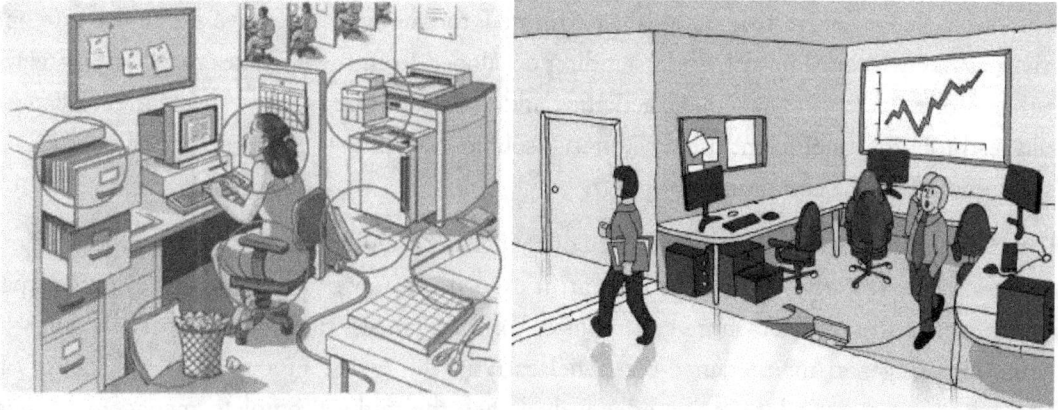

Lock Out Tag Out:

Lock Out-Tag Out (LOTO) or Lock and Tag is a safety procedure which is used across the industry to ensure safety when undertaking maintenance activity on machineries or electrical infrastructure. The LOTO ensure that machine/electrical infrastructure is not able to be energized again prior to the completion of maintenance or repair work.

It requires that hazardous energy sources be "isolated and rendered inoperative" before work is started on the equipment in question. The isolated power sources are then locked and a tag is placed on the lock identifying the worker who placed it. The worker then holds the key for the lock ensuring that only he or she can remove the lock and start the machine. This prevents accidental startup of a machine while it is in a hazardous state or while a worker is in direct contact with it.

Disconnecting or making safe the equipment involves the removal of all energy sources and is known as isolation. The steps necessary to isolate equipment are often documented in an isolation procedure or a lockout tag out procedure. The isolation procedure generally includes the following tasks:

> Announce shut off

> Identify the energy source(s)

> Isolate the energy source(s)

> ➢ Lock and tag the energy source(s)

> ➢ Prove that the equipment isolation is effective

The locking ensures that other person can't dis-isolate the point and tagging lets others know not to de-isolate the device.

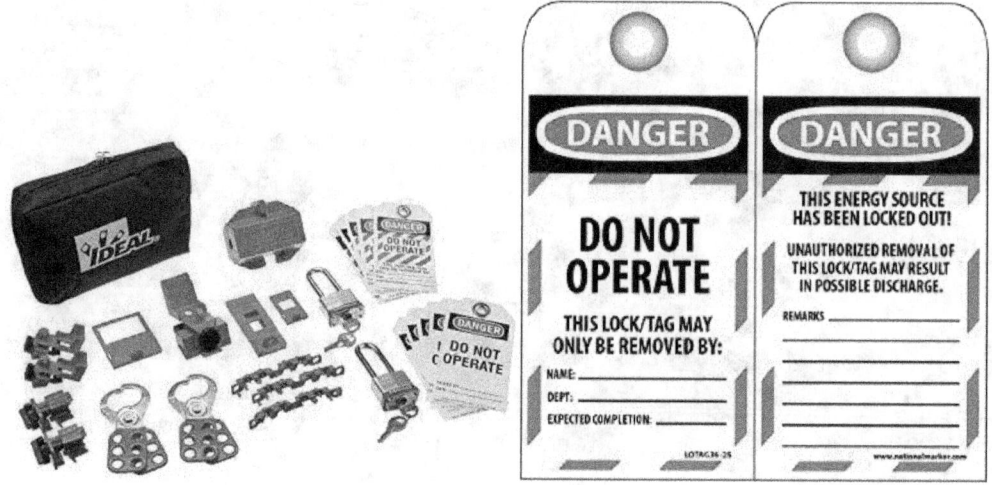

The National Electric Code states that a safety/service disconnect must be installed within sight of serviceable equipment. The safety disconnect ensures the equipment can be isolated and there is less chance of someone turning the power back on if they can see the work going on. These safety disconnects usually have multiple places for locks so more than one person can work on equipment safely.

Group lockout

When two or more people are working on the same or different parts of a larger overall system, there must be multiple holes to lock the device. To expand the number of available holes, the lockout device is secured with a folding scissors clamp that has many pairs of padlock holes capable of keeping it closed. Each worker applies their own padlock to the clamp. The locked-out machinery cannot be activated until all workers have removed their padlocks from the clamp.

In the United States a lock selected by color, shape or size, such as a red padlock, is used to designate a standard safety device, locking and securing hazardous energy. No two keys or locks should ever be the same. A person's lock and tag must only be removed by the individual who installed the lock and tag unless removal is accomplished under the direction of the employer. Employer procedures and training for such removal must have been developed, documented and incorporated into the employer energy control program.

The Lock out- Tag out kid is radially available in market and is a must to have kit for every facility.

The five security steps

According to the European standard EN 50110-1 and Indian Standard IS 5216 1982 part I and II, the safety procedure before working on electric equipment comprises the following five steps:

- disconnect completely;
- secure against re-connection;
- verify that the installation is dead;
- carry out earthing and short-circuiting; and
- provide protection against adjacent live parts.

Lock out and Tag out can be used on:

- Mechanical equipment's/ Machineries
- Electrical Infrastructure
- Fire Hydrant system

Working at Height

Any kind of accident involving men and machineries affects the work, organization, personal life of the individual and the family. Working at a height is a high risk job. One has to take utmost care and use appropriate PPE when working at height. It takes one mistake to turn a routine work task into a fatality. Falls are debilitating and are deadly. One must be prepared to protect your employees each and every time they could be exposed. Here are few tips to consider if your employees work at heights.

a) Use Railing

When you can, use railing. Passive protection is the easiest way to keep your workers safe in order to achieve compliance because there is nothing that they need to actually do to keep themselves safe (other than stay within the rails…and if your employees are climbing outside of protective rails, you've got bigger problems to address!). There are railing system for almost every style of rooftop like non-penetrating railing for flat or low-slope roofs, parapet mounted railing, metal roof railing, and more. Pre-fabricated railings can be permanently affixed or portable to suit your needs. Regardless of which type you use, once in place, you'll find rails to be the easiest fall protection system to use.

b)   Select the Proper PPE

If you're going to use Personal Fall Arrest Systems (PFAS), you need to ensure you're choosing the proper equipment. All full-body harnesses that meet ANSI (American National Standard Institute) standards will perform the same. There are various kind of body harness is available, which can be defined as full party or part body harness. The harnesses are available with various work related features such as extra D-rings, fireproof material or arc-safe design. If you have workers welding at heights, then a standard nylon harness is not abdicable.

Picture Courtesy: https://gravitec.com/equipment/safety-posters/full-body-harness-safety-poster-

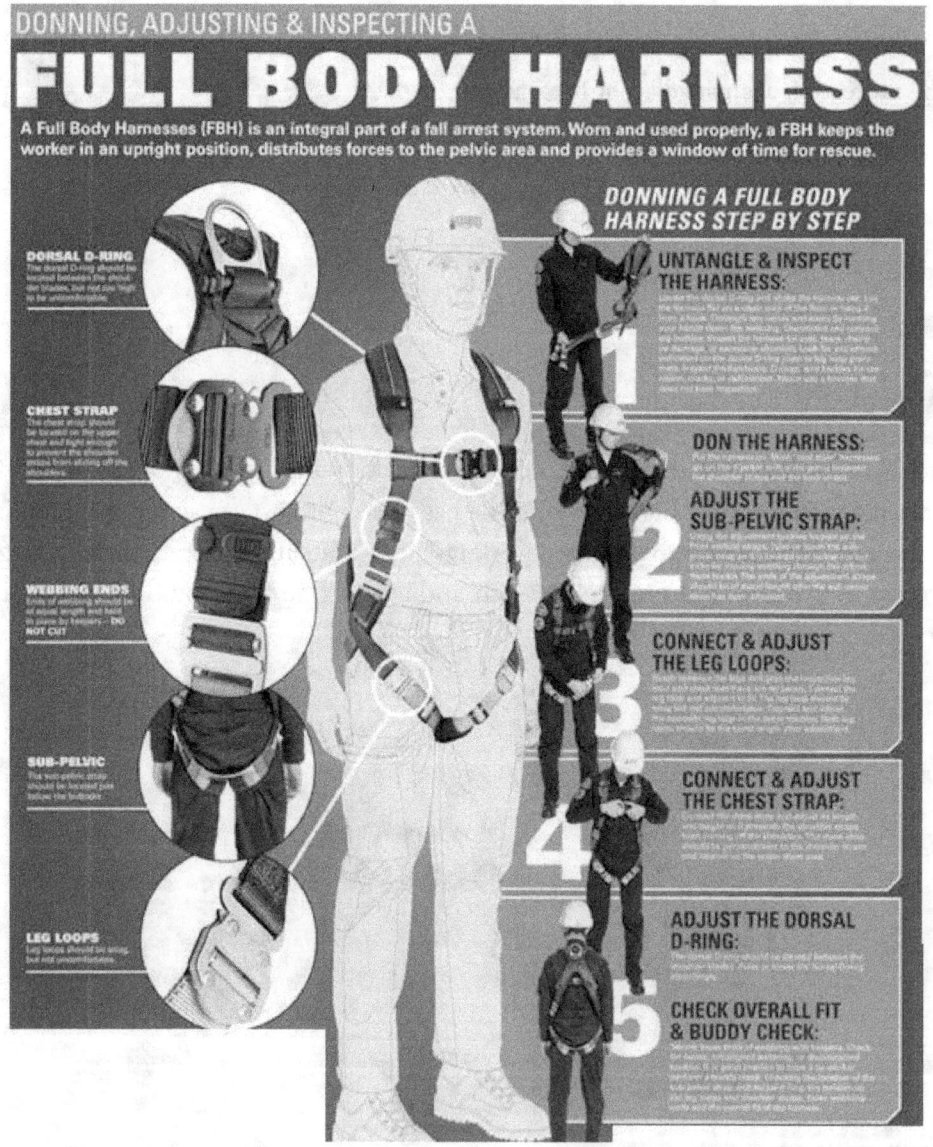

Lanyards need to be properly selected as well. Depending on the height at which you are working, a 6' lanyard with a deceleration device will not protect your worker. Instead, a retractable lanyard may be necessary.

c) **Inspect Your PPE**

Supervisor is required to inspect these PPE's periodically and if finds any kind of damage to PPE he/she should ensure that the PPE is completely damaged (cut in pieces) and discard so that no one should be able to use it.

It is also the responsibility of worker to inspect the PPE before donning it and getting on job, if fail to inspect it could fail anytime.

d) **Ensure the Selection of Acceptable Anchor Point**

The anchor point is an important part of working at height PPE, because it must support not only the weight of the person attached but 5000 lbs. per person attached (or a factor of 2. To see an example of a compliant, and easy to install anchor point take a look at the Weightanka deadweight anchor. Short of that, you're going to need some documentation and/or an engineer's approval to use something as an anchor point.

e) **Use Ladders Properly**

Ladders lie at the source of many industrial and workplace accidents simply because we all take its usage for granted. When improperly used, they're REALLY dangerous. First, make sure that ladders are the best way to do what you're doing, then make sure your employees know how to properly use them. 3' extension, 4:1 ratio, 3 points of contact, and secured.

If you are using a fixed ladder, make sure that it is protected by a ladder safety guard.

➤   Hand Tools, when working at height there are always chances of tool drops. Imagine a tool drop from height directly on someone's head. This may result into fatal accident. In order to avoid any kind of accident because of tool drop, one should ensure that the hand tools used are tied with lanyards. It's critical that all elevated tools and materials are always secured in addition, once the job is finished, workers should double check to make certain no loose objects are left behind.

Also ensure that a restricted area is established beneath suspended objects or work being done at height.

Personal Protective Equipment (PPE)

PPE is wearable equipment which is worn by worker to safeguard against any injury or health related issue caused by the work he/she is performing. A PPE is protective clothing, helmets, mask, goggles, gloves of various kind, shoes, body harness etc.

The PPE is part of contingency plan to reduce residual risk and reduce employee exposure to hazards when engineering controls and administrative controls are not feasible or effective to reduce these risks to acceptable levels. PPE is needed when there are hazards present. PPE has the serious limitation that it does not eliminate the hazard at the source and may result in employees being exposed to the hazard if the equipment fails.

Any item of PPE imposes a barrier between the wearer/user and the working environment. This can create additional strains on the wearer; impair their ability to carry out their work and create significant levels of discomfort. Any of these can discourage wearers from using PPE correctly, therefore placing them at risk of injury, ill-health or, under extreme circumstances, death. PPE's of good ergonomic design can help to minimize these barriers and can therefore help to ensure safe and healthy working conditions through the correct use of PPE.

Helmet

Industrial safety helmets

According to EN 397:1995, the most common and basic form of PPE aimed at protecting an employee's head is an industrial safety helmet. Regardless of the differences in their structural protection, these type of helmets will feature the following components: shell, harness and headband.

Figure 1: Construction of an industrial safety helmet. 1 – shell, 2 – harness, 3 – harness fixing, 4 – headband, 5 – sweatband, 6 – peak, 7 – chinstrap.

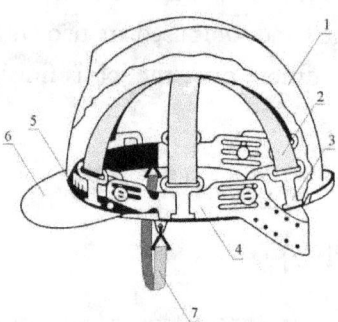

The helmet's shell is the rigid outer section of the helmet, and is usually made from polyethylene, ABS (Acrylonitrile Butadiene Styrene – a thermoplastic) or fiber glass hardened with polyester resins The basic function of the shell is to provide protection by reducing the force of a falling object striking or impacting on the users head. Depending on its design, the shell can have a peak, a brim or a rain gutter, ventilation openings, or attachment devices for eye and face protection and ear protectors.

Proper use of the Safety helmets

The most important rules of proper use of safety helmets include:

➢ Prior to using, a helmet must be fitted to the user's head by proper adjustment: of the headband, height of wearing and the length of the chinstrap (if it is present).

➢ Helmet must be withdrawn from service if it was exposed to a strong impact or shows signs of damage.

➢ Interior elements of a safety helmet must be regularly inspected (harness, headband, sweatband) as they are exposed to sweat, dust, etc. These factors cause an accelerated degradation of the materials of which the helmet components are made. Parts inside the helmet shall be replaced as often as required by the manufacturer and every time any damage is detected during inspection. In the case of doubts, interior elements shall be replaced or the entire helmet substituted by a new one.

➤ Helmet should be withdrawn from service if its expiry date, specified by the manufacturer in the operating manual, has passed.

➤ Helmet shall be stored in compliance with the conditions specified by the manufacturer, which pose no threat of losing its safety parameters (far from heat sources, direct solar radiation, etc.).

➤ The construction of the helmet must not be modified by users, no stickers shall be attached to the shell nor shall it be painted, etc.

➤ The maintenance of the helmet shall be conducted using methods recommended by the manufacturer. Usually helmets can be cleaned with mild detergents and warm water (not hotter than 45°C).

Color Coding for Safety Helmets

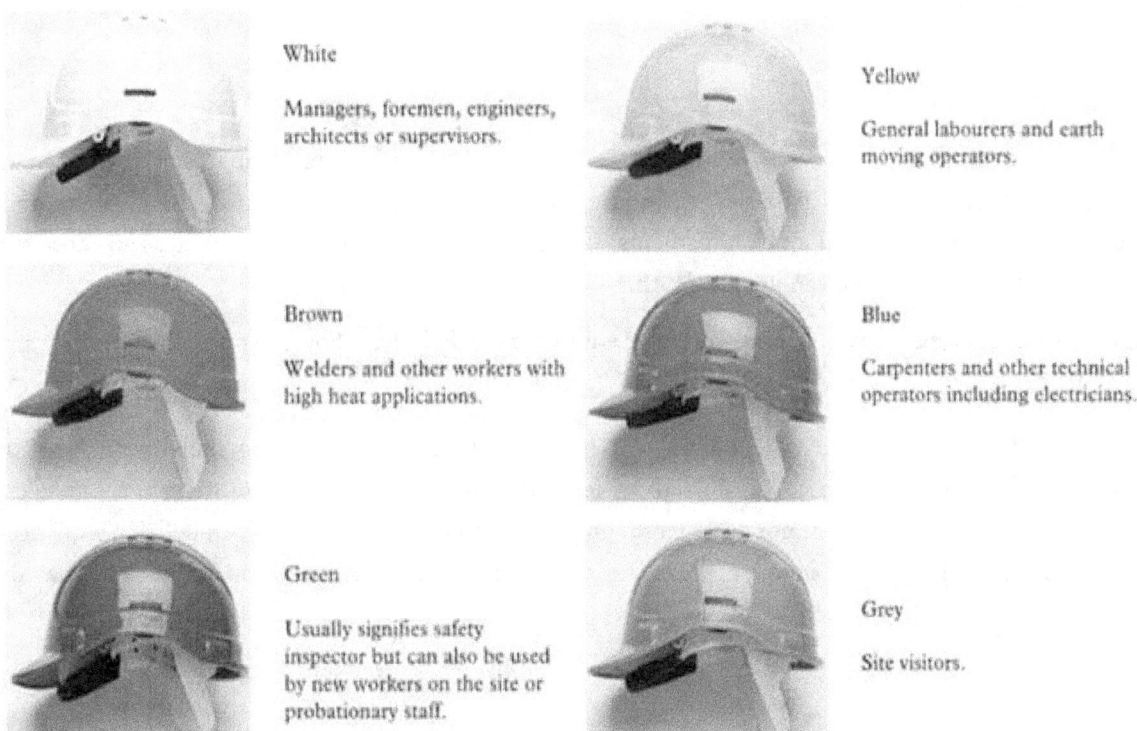

White

Managers, foremen, engineers, architects or supervisors.

Yellow

General labourers and earth moving operators.

Brown

Welders and other workers with high heat applications.

Blue

Carpenters and other technical operators including electricians.

Green

Usually signifies safety inspector but can also be used by new workers on the site or probationary staff.

Grey

Site visitors.

Life of Helmet

➤ The service life begins when the helmet is placed into service. Storage – the maximum shelf life on an unused JSP industrial helmet is five years, as long as it has been stored in reasonable conditions (i.e. not exposed to extremes of temperature, moisture or light).

➤ Working Life of Safety Helmets. … This standard recommends that in general terms the helmet should be replaced every three years based upon industry testing, however to maintain the helmet, the complete head harness insert must be replaced every two years.

Safety Shoes

Safety shoes to be used should be in compliance with IS 15298-Part 2, 2011 ISO 20345: 2004 and IS 5852 for Toe Caps for Footwear.

The life of safety shoes depends on your work environment, your shoes could last shorter or longer. Generally, work safety toe shoes will last between six and twelve months in the average work environment.

Protective Clothing

Protective clothing is any clothing specifically designed, treated or fabricated to protect personnel from hazards that are caused by extreme environmental conditions, or a dangerous work environment.

Some protective clothing is designed to protect the workers from the working environment due to infection or pollution. Protective clothing or any protective equipment is often referred to as personal protective equipment (PPE).

Most of the time protective clothing are prepared from fire resistant material and provided with reflective strips (easy to identify a person working in dark).

Protective Mask

Mask protect's the wearer's face and eyes and prevent the breathing of contaminated air with chemical and/ or biological agents. Solid and liquid particles, including nanoparticles, i.e. dusts, fumes, mists, fibers, radioactive particulates as well as vapors, gases and micro-organisms encountered in workplace's atmospheres can cause significant hazards to health or, in extreme cases, can lead to death.

There are two general types of respiratory protective equipment (RPE), based on the principle by which protection is provided to the user. The two types are the following:

> Respirators (filtering equipment) i.e.: filter, gas filter, combined filter, filtering half-mask. Respirators are designed to filter out or clean contaminated air from the workplace atmosphere before it is inhaled by the respirator wearer. Respirators are not designed to be used in atmospheres with oxygen deficiency (concentration of oxygen is below 19%) or where the concentration of unknown contaminants has not been evaluated.

> Breathing apparatus (isolating equipment) SCBA- Self Contained Breathing Apparatus (open-circuit and closed circuit), compressed line breathing apparatus. Breathing apparatuses deliver breathable air from an independent source (compressed air vessels, compressed line) to the user. Breathing apparatuses are designed to use in atmosphere with oxygen deficiency (concentration of oxygen is below 19%).

It is of key importance that any RPE that is provided to the workers meet the basic requirements specified in the Directive 89/686/EEC concerning Personal Protective Equipment (PPE) and that they are CE marked.

Hand & Finger Safety

Hands a god's gift to mankind. It is such an important tool that one can undertake any kind of tasks from their hands. Hence protection of hands from any kind of injury, amputation or damage is very important as there are no spare parts available for our hands.

Hand injury normally happens due to negligence during work. A hand and finger needs to be protected from smash, scratch, burn freeze. Many of the injuries people sustain may be slight, often barely noticeable. But too frequently, the results to the hands could be worse, even catastrophic.

Types of gloves

> **Insulating gloves: Used for working on electrical infrastructure.**

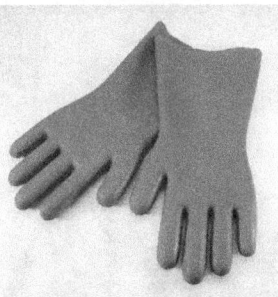

> **Cut Resistant Gloves:**

➤ **Welding Gloves**

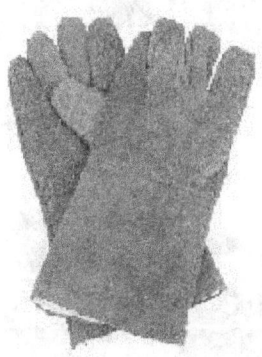

➤ **Chemical protective gloves: Used while handling chemicals.**

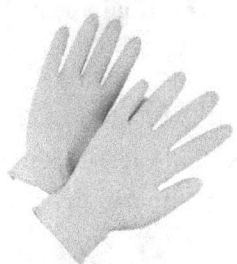

➤ **Impact Resistance Gloves:**

➤ **Handyman Gloves:**

Protective Glasses (Eye wear)

Protective eyewear that usually enclose or protect the eye area in order to prevent particulates, infectious fluids, or chemicals from striking the eyes.

Types of eye wear: Sunglasses, Chemical fume glasses, welding glasses, Grinding Glass, Laser Glasses.

Confined Space

It is a space which has limited entry and egress, almost no ventilation and is not suitable for human inhabitants. An example is the interior of a storage tank, it can be septic tank, water tank, diesel storage or chemical storage tank etc. occasionally entered by maintenance workers but not intended for human occupancy.

Working in confined space is highly hazardous job. Workers will always be at risk while working in confined space from noxious fumes, low oxygen levels, or a risk of fire. Other dangers may include flooding/drowning or asphyxiation from some other source such as dust, grain or other contaminant.

Precaution while working in confined space:

➢ A competent person should undertake risk assessment of the job and have mitigation plan in place.

➢ Personnel entering confined space should be trained.

➢ Before entering confined space one has to ensure that the oxygen level is at acceptable level i.e. 19.5% and 23.5% as suggested by OSHA.

➢ Lockout & Tag out all the power supply to the tank and its associated equipment's.

➢ The tank should be ventilated and adequately purged from gases before entering.

➢ Check the size of man hole (entrance of the tank) to ensure the worker can easily get in with all PPE or evacuated easily in case of emergency.

➢ Ensure to open all the openings of the tank for better ventilation.

➢ Test the air in tank for toxic or flammable gas before allowing entry.

➢ Ensure only special tools are used inside the confined space. Use only 24-volt lamp, non-sparking tools only to be used.

➢ Person should be wearing right kind of PPE for the task.

➤ In case painting job is undertaken ensure continuous extraction of paint fume and person should be wearing mask.

➤ Person should wear body harness and the lanyard should be attended by supervisor. In case of emergency the worker can pull the cable or signal to the supervisor for evacuation.

➤ Ensure emergency plan is in place before starting the work. Following items should be made available at workplace in sufficient:

  ➤ Approved breathing apparatus

  ➤ Suitable reviving apparatus

  ➤ Vessels containing oxygen

  ➤ Safety Harness and ropes

  ➤ An audio visual alarm to alert people outside the confined space.

  ➤ Stretcher

  ➤ First Aid kit

SIGNAGE FOR Confined space

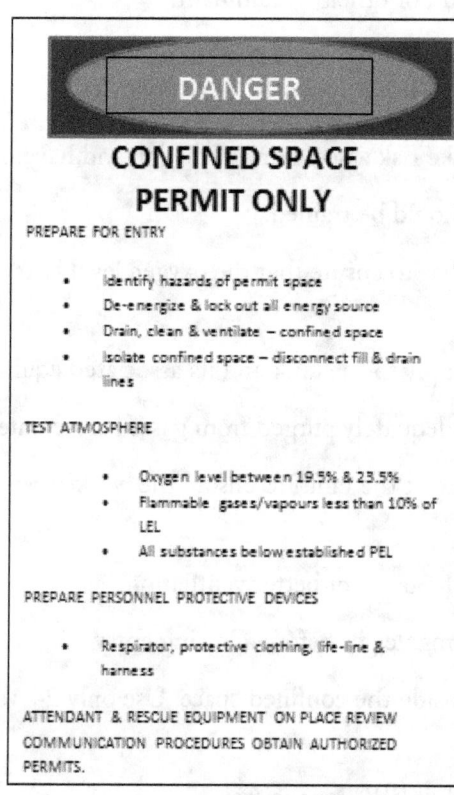

Warning Sign to be placed at opening of confined space

Work Permit Form for Confined Space

CONFINED SPACE ENTRY PERMIT

Confined Space Location/Description/ID Number: Date:

Purpose of Entry:

| Time In:_____ | | | Permit Canceled Time: _____ |
| Time Out:_____ | | | Reason Permit Canceled: _____ |

Supervisor:_____

Rescue and Emergency Services-

Hazards of Confined Space	Yes	No	Special Requirements		YES	NO
Oxygen deficiency			Hot Work Permit Required			
Combustible gas/vapor			Lockout/Tag out			
Combustible dust			Lines broken, capped, or blanked			
Carbon Monoxide			Purge-flush and vent			
Hydrogen Sulfide			Secure Area-Post and Flag			
Toxic gas/vapor			Ventilation			
Toxic fumes			Other- List:			
Skin-chemical hazards				Special Equipment		
Electrical hazard			Breathing apparatus-respirator			
Mechanical hazard			Escape harness required			
Engulfment hazard			Tripod emergency escape unit			

(Contd.)

			Lifeline		
Entrapment hazard			Lifeline		
Thermal hazard			Lighting (explosive proof/low voltage)		
Slip or fall hazard			PPE-goggles, gloves, clothing, etc.		
			Fire Extinguisher		
Communication Procedures:					

DO NOT ENTER IF PERMISSABLE ENTRY			Test Start and Stop Time:		
LEVELS ARE EXCEEDED			Start		Stop
	Permissible Entry Level				
% of Oxygen	19.5% to 23.5%				
% of LEL	Less than 10%				
Carbon Monoxide	35PPM (8 hr.)				
Hydrogen Sulfide	10 PPM (8 hr.)				
Other					

Names(s) of Person(s) testing:_____

_____

Test Instrument(s) used-Include Name, Model, Serial Number and Date Last Calibrated:

_____

CFM-Ventilation	Size-Cubic Feet	Pre Entry Time		[ ] Central Notified	Time Notified:
				Before Entrance	
				[ ] Central Notified	Time Notified:
				After Entrance	
Authorized Entrants				Authorized Attendants	
				_____	
_____				_____	
				_____	
_____				_____	
				_____	
_____				_____	
	PERMIT AUTHORIZATION				

I Certify that all actions and conditions necessary for safe entry have been performed.

Name-			
Signature:			
Date:		Time:	

Entry Procedure Checklist: Complete the following steps before, during, and after a confined space entry:
Step 1 Obtain a Permit-confined Space Entry Form from Program Coordinator.
Step 2 Notify Supervisor before the Confined Space Entry
Step 3 Verify Confined Space Meter has been calibrated and is in working order.
Step 4 Complete the top portion of the Permit-Confined Space Entry Form
Step 5 Ensure all rescue equipment (e.g. tripod, body-belt, lanyard) is in place prior to entry
Step 6 Monitor the confined space with the MSA 4-Gas Detector prior to entry. The entrant and attendant should sign the permit authorization section on the bottom of the permit to ensure all actions and conditions necessary for safe entry have been performed.
Step 8 Employee can enter the confined space once Step 7 is completed. The entrant and attendant should complete the Hazards of Confined Spaces and Special Requirements Section of the Permit-Confined Space Entry Form once the employee is within the confined space. The entrant should also gather the % Oxygen, % Explosive Gases, Carbon Monoxide, and Hydrogen Sulfide readings and communicate them to the attendant to place on the Permit Form.
Step 9 The attendant should maintain constant communication with the entrant until the entrant has exited the confined space.
Step 10 The attendant should contact Supervisor once the entrant has exited the confined space.
Step 11 The Permit-Confined Space Entry Form should be given to program coordinator, to file in the Confined Space Records.

Ladder Safety: There are many incidents where fall from ladder has resulted in serious injuries. One has to be very cautious when working on ladder. It is to be ensured that the ladders should be long enough to extend at least 3 feet above the landing. And they should be tied off. They should also be set at a proper angle. A 1 to 4 pitch is recommended. One foot out for each four feet of height.

And no matter how convenient it might seem, don't try to carry tools and materials up or down the ladder. Use a hand line to haul them up or down.

Tips for ladder safety

➤ Carefully inspect the ladder for defects, checking for cracks, corrosion and that bolts and rivets are secure. Tag and remove unsafe ladders from service.

➤ Make sure the ladder's feet work properly and have slip-resistant pads.

➢ Use a fiberglass ladder if there is any chance of contact with electricity.

➢ When setting the ladder, look for a safe location with firm, level footing and rigid support for the top of the ladder. Be sure to set it at an angle per the manufacturer's guidance.

➢ When climbing off a ladder at an upper level, make sure the ladder extends 3 feet above the landing.

➢ When climbing the ladder, use three points of contact — keep 1 hand and both feet or both hands and 1 foot in contact with the ladder at all times.

➢ Never carry any load that could cause you to lose balance.

➢ Never stand on top of a ladder.

➢ Don't pull, lean, stretch or make sudden moves on a ladder that could cause it to tip over. A scaffold or other safe working surface may be a better choice for your task.

➢ Avoid setting the ladder near exit doors, near the path of pedestrian or vehicular traffic.

Ladder Maintenance & Upkeep

Most of the time people have tendency to take things like ladders for granted. It is easy for people in our country to neglect ladder maintenance and only provide care when obvious defects appear. Negligence in ladder maintenance may lead to fatal accidents.

Failing to maintain ladders can compromise their integrity and even increase the risk of injury for users. Hence, to avoid such unfortunate scenarios, it pays to be vigilant and ensure ladder safety by performing regular care and maintenance.

Whether you own extension, platform step, or attic ladders, here are some tips you should keep in mind.

Maintenance tips for ladder

➢ Before and after every use, ensure the equipment is clean and free of oil, mud, grease or any other substance that may make the steps or rungs slippery.

➢ Ensure not to drop the ladder. Ladders that receive a heavy blow may become damaged, have loosened steps or rungs and deformed stiles. Should this happen, inspect the equipment thoroughly before using it.

➢ Check for missing or loose rungs, damaged feet, faulty spreaders, rot or decay in wooden equipment, twisted rails, splintered surfaces, and missing identification labels.

➢ Any defective equipment must be suitably labelled or taken out of service.

Chemical Handling: Improperly storing, handling and mixing of chemicals can cause significant safety and environmental risks. One should ensure that before working with chemicals proper control such as spill kits are used and personal protective equipment's (PPE) such as gloves, goggles are used and the people involved in chemical handling are trained and aware of nearest eye wash and safety showers.

Also ensure that the safety data sheet (SDS) also known as MSDS (Material Safety Data Sheet) (of chemical is reviewed by the handlers and they are aware of potential hazards involved. If work with chemicals is to be undertaken in confined space, allow only trained workers with a work permit to perform the task.

Should spills of hazardous chemical happens take emergency response action as indicated in emergency response plan. Some spills may require government reporting so one should be familiar with government compliance requirement also.

Spill Kits are compilation of absorbent materials, cleaners, and chemical neutralizers - used to contain accidental spills in an industrial setting.

Design of Spill Kits

A spill kit should normally contain following items:

➢ Biohazard bags

➢ Disinfectants

➢ Respirators or masks

➢ Gloves

➢ Eye protection

➢ Shoe covers

➢ Sharps containers and/or tools used in picking up broken glass or/and sharp objects

➢ Absorbent material specifically designed for handling the type of hazard on site

In addition, the absorbents are flexible when it comes to their absorbing potential and design, largely based on the kind of liquid that needs to be stored, sorbents could feature one of the following designs:

➢ **Universal or general purpose:** This is designed for handling a wide range of no corrosive type of spilled liquids

➢ **Oil and fuel:** This is ideal for spills that are oil based

> ➤ **HAZCHEM**: To handle aggressive chemicals.

Chemical Compatibility: There are various kinds of chemicals which are incompatible with each other. Storing of incompatible chemicals in too close proximity to each other may lead to release of toxic, corrosive and flammable vapors and liquids. This may further cause excessive heat generation and even explosion. With some chemicals, exposure to air or water might trigger such reactions.

Hazard Symbols

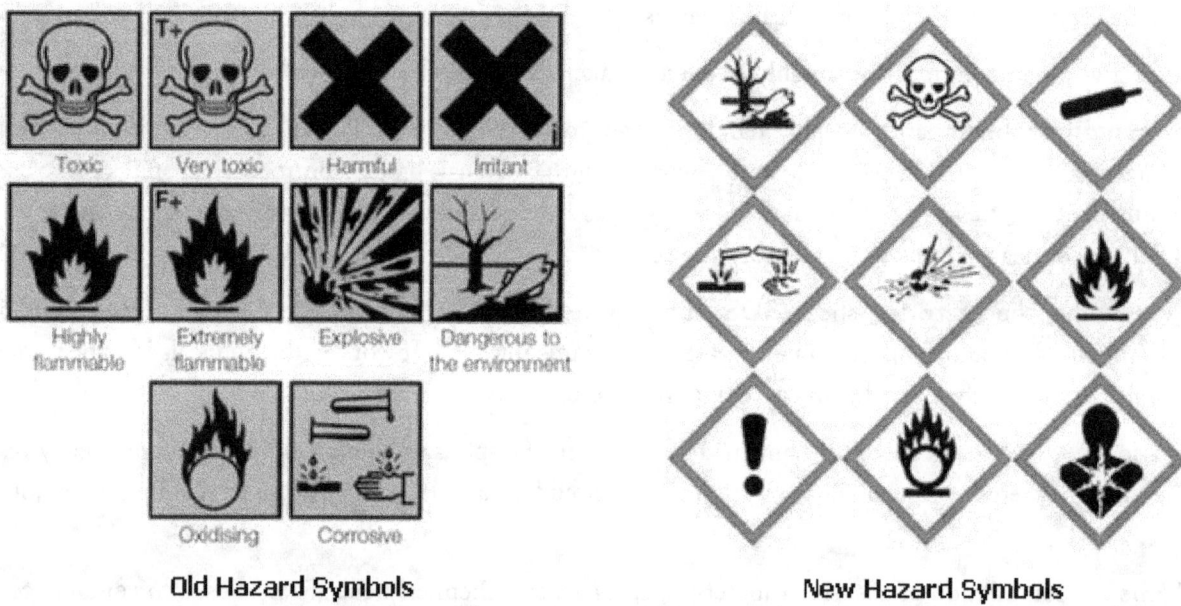

Old Hazard Symbols   New Hazard Symbols

Slip, Trips & Falls: People must be able to move around the workplace safely. Slips, trips and falls are among the most common causes of accidents and injury at work.

1. **Common** slip, trip and fall hazards are caused by

   > ➤ poor lighting

   > ➤ trailing cables

   > ➤ unsuitable floor coverings

   > ➤ uneven or damaged floor surfaces

   > ➤ contaminated floor surfaces, for example liquid or grease

   > ➤ poor housekeeping, for example tripping or falling over something left in a walk way.

   > ➤ Wet floor

Slip, trip and fall safety precautions: Both slips and trips results because of unintended or unexpected change in the contact between the feet and the ground or walking surface. This indicates that good housekeeping, quality of walking surfaces (flooring should be anti-skid), selection of proper footwear (anti-skid), and appropriate pace of walking are critical for preventing fall incidents. There are two important factor affecting slips and trips and they area:

> ➤ **Housekeeping**

   Good housekeeping is the first and the most important (fundamental) level of preventing falls due to slips and trips. It includes:

- cleaning all spills immediately

- marking spills and wet areas (signage, barricading)

- mopping or sweeping debris from floors

- removing obstacles from walkways and always keeping walkways free of clutter

- securing (tacking, taping, etc.) mats, rugs and carpets that do not lay flat

- always closing file cabinet or storage drawers

- covering cables that cross walkways

- keeping working areas and walkways well lit

- replacing used light bulbs and faulty switches

- **Flooring**

  To prevent slip and trips, replacing of floorings with anti-skid flooring is always advisable. Recoating of floors, installing mats, pressure-sensitive abrasive strips or abrasive-filled paint-on coating and metal or synthetic decking can further improve safety and reduce risk of falling. However, it is critical to remember that high-tech flooring requires good housekeeping as much as any other flooring. In addition, resilient, non-slippery flooring prevents or reduces foot fatigue and contributes to slip prevention measures.

- **Footwear**

  In workplaces where floors may be oily or wet or where workers spend considerable time outdoors, prevention of fall incidents should focus on selecting proper footwear. Since there is no footwear with anti-slip properties for every condition, consultation with manufacturers' is highly recommended.

  Properly fitting footwear increases comfort and prevents fatigue which, in turn, improves safety for the employee. For more information on footwear visit the OSH Answers document on Safety Footwear.

Management of Change (MOC)

Management of Change, or MOC, is a process used to ensure that safety, health and environmental risks are controlled when an organization makes changes in their facilities, documentation, process, personnel, or operations.

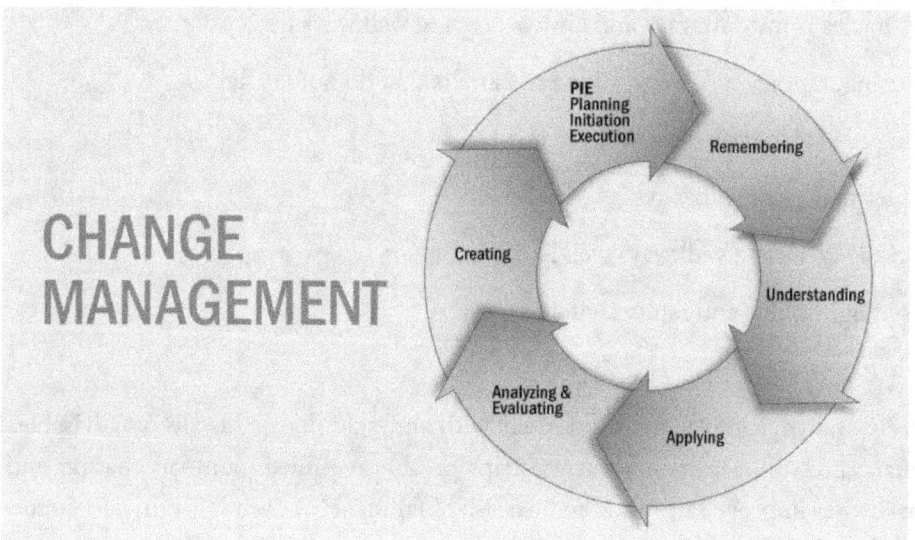

Let's work with this simple definition: MOC is a process for preventing or mitigating business losses including degradation of safety, health or environment as the result of changes made to how you construct, operate, manage, or repair your facility or your processes.

The objective of a Management of Change (MOC) program is to ensure all changes to a process are properly reviewed and hazards introduced by the change are identified, analyzed, and controlled prior to resuming operation. MOC often seems deceptively simple in concept but can be one of the most difficult elements of Process Safety Management to implement effectively.

The MOC must include descriptions of the technical basis for the change, impact on safety and health, modifications to operating procedures, the time period necessary for the change, and appropriate authorizations. The MOC section of the standard also states that any employee who will be affected by the change must be informed and appropriately trained. According to OSHA, common areas of non-compliance with MOC requirements include managing changes to equipment design and operating procedures, regular maintenance and repair to facilities, and documenting time limits for temporary changes.

Changes in Inspection and Test and Maintenance Procedures: Whenever operators change inspection and maintenance procedures, including changes to preventative maintenance and equipment repair, they need to utilize MOC, since these changes can affect the risk levels of the associated process or equipment. Examples; For example you have planned to change the cleaning process of HVAC condenser, and the process you are planning will be deviation from the previous process will require an MOC. You can consider another example of change in cleaning chemical. This will also require an MOC.

Changes in Facilities: As per the OSHA's Process Safety Management (PSM) standard MOC is required for any kind of change or modification in facilities. This requirement is triggered whenever an existing structure is modified, or a newly installed facility structure will be located within or near a PSM-covered process. Potential examples include expansion of Data Center or server room, or creation of an additional business unit functional area.

MOC Forms Sample

Change Request	
Project:	**Date:**
Change Requestor:	**Change No:**
Change Category (Check all that apply): ☐ Schedule ☐ Cost ☐ Scope ☐ Requirements/Deliverables ☐ Testing/Quality ☐ Resources	
Does this Change Affect (Check all that apply): ☐ Corrective Action ☐ Preventative Action ☐ Defect Repair ☐ Updates ☐ Other	
Describe the Change Being Requested:	
Describe the Reason for the Change:	
Describe all Alternatives Considered:	
Describe any Technical Changes Required to Implement this Change:	
Describe Risks to be Considered for this Change:	
Estimate Resources and Costs Needed to Implement this Change:	
Describe the Implications to Quality:	
Disposition: ☐ Approve ☐ Reject ☐ Defer	
Justification of Approval, Rejection, or Deferral:	

Change Board Approval:		
Name	**Signature**	**Date**

Driving Safety

All the corporates use either owned vehicle or leased vehicle for transportation of their employees and management. And the safety of employees on road always remains in hand of the driver. It is always advisable to ensure the drivers are medically and mentally fit. The drivers are required to be soft in nature and disciplined on the road.

It is always advisable to ensure that the drivers are defensive driving trained and a refreshment course is conducted every year. Apart from above as a facility/transport manager one should ensure that daily briefing and weekly checks are undertaken for drivers and vehicles.

Driving an automobile is one of the potentially most dangerous things people do, but you can prevent certain problems if you know how to check your car before driving. Visual inspections may prevent an accident caused by a blown tire, and many other potential hazards. Road accidents can lead to fatal injury or death. One has to be cautious before starting their road trip and while driving. The visual inspection to be undertaken is as appended below:

➢ Check under the car for obvious leaks. Driving with leaking fluid may cause failure of the steering, brakes or radiator.

➢ Check the tires for proper inflation and any obvious damage or signs of excessive wear. In a worst-case scenario, a blown tire could cause you to crash.

➢ Check all lights are operational.

➢ Check a first aid kit is always available in the car.

➢ Ensure a fire extinguisher is placed in car and is maintained and operational.

➢ Check oil level and coolant level in car, it should of prescribed level.

➢ Check the gauge's every time you start your car. Check the engine temperature gauge after the engine has had time to warm.

➢ Check fuel level.

➢ Has the battery tested before a trip?

➢ Check your air filter before a long trip, as it can affect fuel efficiency and engine performance.

➢ Make sure the spare tire is inflated and serviceable and the jack is present. It's a good idea to check them periodically even if you aren't going on a long trip.

➢ Ensure seat belts are in working condition.

Precautions while driving a vehicle

➢ Wear seat belts

➢ Don't text or talk on mobile

➢ Be aware of what other drivers around you are doing, and expect the unexpected.

➢ Assume other motorists will do something crazy, and always be prepared to avoid it.

➢ Keep a 2-second cushion between you and the car in front of you. Make that 4 seconds if the weather is bad.

➢ Follow speed limit.

➢ Don't get your vehicle in a flooded road

➢ Stay Alert - Actively pay attention to your actions and those of the drivers around you when you are driving.

➢ Avoid Assumptions - Don't make the mistake of assuming that other drivers are going to do or what you think they should do.

- ➤ Use Turn Signals - While you can't depend on others always signaling their intentions when driving, you can certainly control whether or not they have realistic expectations for your actions. Always use your turn signals in advance of making a lane change or turning.

- ➤ Follow Traffic Signals - Pay close attention to and obey stop signs and traffic lights.

- ➤ Respect Yellow Lights - Remember that the intent of a yellow light is to notify drivers to slow down and prepare to stop. A yellow traffic signal should not be viewed as a sign to step on the gas to rush through an intersection before the light turns red.

- ➤ Come to a Complete Stop - When you see a stop sign or a red light, it's important to bring your vehicle to a complete stop, even if you think no other vehicles are coming.

- ➤ Make Adjustments for Weather-When the weather is less than perfect, such as rainy, snowy, or foggy conditions, use extra precautions when driving and follow guidelines for staying safe in the particular situation you are facing.

- ➤ Exercise Patience - Many accidents are caused by impatient drivers who are rushing to get from point A to point B. While time is certainly a valid consideration when traveling, safety is even more important. After all, if you are involved in an accident you'll certainly experience more of a challenge arriving at your destination on time than if you simply exhibit patience while driving.

- ➤ Be Predictable - Don't make sudden stops or lane changes. Instead, take care to ensure that other drivers are likely to be able to predict your actions to maximize safety.

- ➤ Never Drive Under the Influence - It's essential to avoid operating a vehicle if you have been drinking, taking certain types of prescription or non-prescription drugs, or are otherwise impaired.

- ➤ Yield Right of Way - When other drivers has the right of way, be sure to yield to them. Also, don't make the mistake of assuming that everyone else will yield to you when they should. Regardless of who has the right to go, yield if it seems that the other driver may not be observing standard practices for yielding.

- ➤ Know Where You Are Going - Plan your travel route ahead of time so that you aren't struggling to figure out where to go while you are operating a motorized vehicle.

- ➤ Respect Stopped Vehicles - When passing vehicles that are stopped on the side of the road, move over to get out of the way if the way is clear for you to change lanes. If changing lanes is not possible, slow down while passing stopped vehicles.

- ➤ Avoid Distractions - Sending text messages isn't the only dangerous distraction that drivers need to avoid while operating a vehicle. Changing CDs, using cell phones, eating, and interacting with passengers are just a few examples of the types of distractions that you should take care to avoid when driving.

- ➤ Use Headlights When Needed - Headlights aren't just necessary at night. When you are driving in the rain or fog, turning on your headlights can play an important role in keeping you - and those around you - safe on the road.

- ➤ Share the Road - Remember that you are not the only driver on the road. An important safety trip that everyone needs to follow is the need to share the road with others graciously, recognizing that all drivers deserve to be treated with respect.

Emergency Response Plan

It is known facts that people get panicked when an emergency situation arises and they don't know what to do in what kind of situation. In view of the same it is important to have an **Emergency Response Plan** in place. The correct actions taken in the initial minutes of an emergency are critical and lifesaving.

A prompt warning to employees to evacuate, shelter or lockdown can save lives. A call for help to public emergency services that provides full and accurate information will help the dispatcher send the right responders and equipment. An employee trained to administer first aid or perform CPR can be lifesaving. Action by employees with knowledge of building/ premises and process systems can help control a leak and minimize damage to the facility and the environment.

Before starting with preparation of an ERP, it is advisable to conduct a risk assessment to identify potential emergency scenarios. The emergency scenario can be failure of equipment, fire, accident, flood, earthquake etc. An understanding of what can happen will enable you to determine resource requirements and to develop plans and procedures to prepare your business.

Every facility should develop and implement an emergency plan for protecting asset, employees, visitors, contractors and others in the facility. This part of the emergency plan is called "protective actions for life safety" and includes building evacuation ("fire drills"), sheltering from severe weather such as flood, cyclone, exterior airborne hazard such as a chemical release and lockdown. Lockdown is protective action when faced with an act of violence.

Life Saving Actions

In case of a hazard such as bomb threat, receipt of suspicious package, fire or chemical spill in building it is responsibility of Safety officer or Facility Manager to ensure that the building is evacuated or people are moved to safe place such as refuge area. In case of threat of violence by miscreants a lockdown should be observed and no-one should be permitted to move out of premises, everyone should hide or barricade themselves from perpetrator.

Protective actions required for life safety

➢ Evacuation

➢ Sheltering

➢ Shelter-In-Place

➢ Lockdown

If you are a tenant in a commercial building, in this case you need to coordinate with the building management team.

Importance of emergency plan

A definite plan to deal with major emergencies is an important element of OH&S programs.

Besides the major benefit of providing guidance during an emergency, developing the plan has other advantages. You may discover unrecognized hazardous conditions that would aggravate an emergency situation and you can work to eliminate them. The planning process may bring to light deficiencies, such as the lack of resources (equipment, trained personnel, supplies), or items that can be rectified before an emergency occurs. In addition, an emergency plan promotes safety awareness and shows the organization's commitment to the safety of workers.

The lack of an emergency plan could lead to severe losses such as multiple casualties and possible financial collapse of the organization.

An attitude of "it can't happen here" may be present. People may not be willing to take the time and effort to examine the problem. However, emergency planning is an important part of company operation.

Since emergencies will occur, preplanning is necessary. An urgent need for rapid decisions, shortage of time, and lack of resources and trained personnel can lead to chaos during an emergency. Time and circumstances in an emergency mean that normal channels of authority and communication cannot be relied upon to function routinely. The stress of the situation can lead to poor judgment resulting in severe losses.

Objective of Emergency Response Plan

An emergency plan specifies procedures for handling sudden or unexpected situations. The objective is to be prepared for:

> Prevent fatalities and injuries.

> Reduce damage to buildings, stock, and equipment.

> Protect the environment and the community.

> Accelerate the resumption of normal operations.

> Reduce or minimize loss of business operations.

The ERP plan development starts with vulnerability assessment. To start with one has should understand what kind of emergencies may occur, then review your risk assessment and answer the following quarries:

> How likely a situation is to occur?

> What means are available to stop or prevent the situation.

> What is necessary for a given situation.

From answers of above quarries and analysis, you can establish appropriate emergency procedures.

At the planning stage, it is important that several groups be asked to participate. Among these groups, the health and safety committee can provide valuable input and a means of wider worker involvement. Appropriate municipal officials should also be consulted since control may be exercised by the local government in major emergencies and additional resources may be available. Communication, training and periodic drills will ensure adequate performance if the plan must be carried out.

Vulnerability assessment

Although emergencies by definition are sudden events, their occurrence can be predicted with some degree of certainty. The first step is to find which hazards pose a threat to any specific enterprise.

When a list of hazards is made, records of past incidents and occupational experience are not the only sources of valuable information. Since major emergencies are rare events, knowledge of both technological (chemical or physical) and natural hazards can be broadened by consulting with fire departments, insurance companies, engineering consultants, and government departments.

Technological and natural hazards

Technological Hazards

> Fire.

> ➢ Explosion.

> ➢ Building collapse.

> ➢ Major structural failure.

> ➢ Spills of flammable liquids.

> ➢ Accidental release of toxic substances.

> ➢ Deliberate release of hazardous biological agents, or toxic chemicals.

> ➢ Other terrorist activities.

> ➢ Exposure to ionizing radiation.

> ➢ Loss of electrical power.

> ➢ Loss of water supply.

> ➢ Loss of communications.

> ➢ Consumption of contaminated water

Areas where flammables, explosives, or chemicals are used or stored should be considered as the most likely place for a technological hazard emergency to occur.

Natural Hazard

➢ Floods

➢ Earthquakes

➢ Tornadoes

➢ Other severe wind storms

➢ Snow or ice storms

➢ Severe extremes in temperature (cold or hot)

➢ Pandemic diseases like influenza

The possibility of one event triggering others must be considered. An explosion can also result into a fire and it further can cause structural failure.

An earthquake might initiate many of the technological events listed above.

Actions to be considered in case of Emergency

Having identified the hazards, the possible major impacts of each should be itemized, such as:

> ➢ Sequential events (for example, a fire after an explosion).

> ➢ Evacuation.

> ➢ Casualties.

> ➢ Damage to plant infrastructure.

> Loss of vital records/documents.

> Damage to equipment.

> Disruption of work.

Based on these events, the required actions are determined. For example:

> Declare emergency.

> Sound the alert.

> Evacuate danger zone.

> Close main shutoffs.

> Call for external aid.

> Initiate rescue operations.

> Attend to casualties.

> Fight fire.

Resources needed to handle emergency

> Medical supplies.

> Auxiliary communication equipment.

> Power generators.

> Respirators.

> Chemical and radiation detection equipment.

> Mobile equipment.

> Emergency protective clothing.

> Fire fighting equipment.

> Ambulance.

> Rescue equipment.

> Trained personnel.

What are the elements of the emergency plan?

The emergency plan includes:

> All possible emergencies, consequences, required actions, written procedures, and the resources available.

> Detailed lists of personnel including their home telephone numbers, their duties and responsibilities.

> Floor plans.

> Large scale maps showing evacuation routes and service conduits (such as gas and water lines).

Since a sizable document will likely result, the plan should provide staff members with written instructions about their particular emergency duties.

The following are examples of the parts of an emergency plan. These elements may not cover every situation in every workplace but serve they are provided as a general guideline when writing a workplace specific plan:

Objective

The objective is a brief summary of the purpose of the plan; that is, to reduce human injury and damage to property and environment in an emergency. It also specifies those staff members who may put the plan into action. The objective identifies clearly who these staff members are since the normal chain of command cannot always be available on short notice. At least one of them must be on the site at all times when the premises are occupied. The extent of authority of these personnel must be clearly indicated.

Organization

Two responsible person to be appointed and trained to act as Emergency coordinator. One person to be trained as a lead and another as a "back-up" coordinator. Personnel available on site during an emergency are the key people in ensuring that prompt and efficient action is taken to minimize loss. In some cases, it may be possible to recall off-duty employees to help, but the critical initial decisions usually must be made immediately.

In the organization specific duties, responsibilities, authority of resources should be clearly defined. Among the responsibilities that must be assigned are:

- Reporting the emergency.
- Activating the emergency plan.
- Assuming overall command.
- Establishing communication.
- Alerting staff.
- Ordering evacuation.
- Alerting external agencies.
- Confirming evacuation is complete.
- Alerting outside population of possible risk.
- Requesting external aid.
- Coordinating activities of various groups.
- Advising relatives of casualties.
- Providing medical aid.
- Ensuring emergency shut offs are closed.
- Sounding the all-clear.
- Advising media.

This list of responsibilities should be completed using the previously developed summary of countermeasures for each emergency situation. In organizations operating on reduced staff during some shifts, some personnel must assume extra responsibilities during emergencies. Sufficient alternates for each responsible position must be named to ensure that someone with authority is available onsite at all times.

External organizations that may be available to assist (with varying response times) include:

➢ Fire departments.

➢ Mobile rescue squads.

➢ Ambulance services.

➢ Police departments.

➢ Telephone companies.

➢ Hospitals.

➢ Utility companies.

➢ Industrial neighbors.

➢ Government agencies.

These organizations should be contacted in the planning stages to discuss each of their roles during an emergency. Mutual aid with other industrial facilities in the area should be explored.

Pre-planned coordination is necessary to avoid conflicting responsibilities. For example, the police, fire department, ambulance service, rescue squad, company fire brigade, and the first aid team may be on the scene simultaneously. A pre-determined chain of command in such a situation is required to avoid organizational difficulties. Under certain circumstances, an outside agency may assume command.

Possible problems in communication have been mentioned in several contexts. Efforts should be made to seek alternate means of communication during an emergency, especially between key personnel such as overall commander, on-scene commander, engineering, fire brigade, medical, rescue, and outside agencies. Depending on the size of the organization and physical layout of the premises, it may be advisable to plan for an emergency control centre with alternate communication facilities. All personnel with alerting or reporting responsibilities must be provided with a current list of telephone numbers and addresses of those people they may have to contact.

Procedures

Many factors determine what procedures are needed in an emergency, such as:

➢ Nature of emergency.

➢ Degree of emergency.

➢ Size of organization.

➢ Capabilities of the organization in an emergency situation.

➢ Immediacy of outside aid.

➢ Physical layout of the premises.

Common elements to be considered in all emergencies include pre-emergency preparation and provisions for alerting and evacuating staff, handling casualties, and for containing the danger.

Natural hazards, such as floods or severe storms, often provide prior warning. The plan should take advantage of such warnings with, for example, instructions on sand bagging, removal of equipment to needed locations, providing alternate sources of power, light or water, extra equipment, and relocation of personnel with special skills. Phased states of alert allow such measures to be initiated in an orderly manner.

The evacuation order is of greatest importance in alerting staff. To avoid confusion, only one type of signal should be used for the evacuation order. Commonly used for this purpose are sirens, fire bells, whistles, flashing lights, paging system announcements, or word-of-mouth in noisy environments. The all-clear signal is less important since time is not such an urgent concern.

The following are "musts":

➢ Identify evacuation routes, alternate means of escape, make these known to all staff; keep the routes unobstructed.

➢ Specify safe locations for staff to gather for head counts to ensure that everyone has left the danger zone. Assign individuals to assist employees with disabilities.

➢ Carry out treatment of the injured and search for the missing simultaneously with efforts to contain the emergency.

➢ Provide alternate sources of medical aid when normal facilities may be in the danger zone.

➢ Ensure the safety of all staff (and/or the general public) first, then deal with the fire or other situation.

Testing and Revision

Completing a comprehensive plan for handling emergencies is a major step toward preventing disasters. However, it is difficult to predict all of the problems that may happen unless the plan is tested. Exercises and drills may be conducted to practice all or critical portions (such as evacuation) of the plan. A thorough and immediate review after each exercise, drill, or after an actual emergency will point out areas that require improvement. Knowledge of individual responsibilities can be evaluated through paper tests or interviews.

The plan should be revised when shortcomings have become known, and should be reviewed at least annually. Changes in plant infrastructure, processes, materials used, and key personnel are occasions for updating the plan.

It should be stressed that provision must be made for the training of both individuals and teams, if they are expected to perform adequately in an emergency. An annual full-scale exercise will help in maintaining a high level of proficiency.

Development of Emergency Response Plan

To start with one has to understand what may go wrong or what may happen, then review your risk assessment. Assess what resources are available for incident stabilization. One should consider available internal and external resources including public emergency services and contractors. Public emergency services include fire departments that may also provide rescue, hazardous materials and emergency medical services. If not provided by your local fire department, these services may be provided by another department, agency or even a private contractor. Reach out to local law enforcement to coordinate planning for security related threats.

Document available resources. Determine whether external resources have the information they would need to handle an emergency. If not, determine what information is required and be sure to document that information in your plan.

Prepare emergency procedures for foreseeable hazards and threats. Review the list of hazards, develop hazard and threat specific procedures.

Prepare a list of employees who will be responsible for handling emergency situations. The document should contain contact number of employees.

The emergency response plans should define the most appropriate action for each hazard to ensure the safety of employees and others within the building.

Warnings and Communications

> There has to be a communication plan in place where it is determined how you will warn building occupants to take protective action.

> Develop protocols and procedures to alert first responders including public emergency services, trained employees and management.

> Identify how and who will communicate with management and employees during and following an emergency.

Roles and Responsibilities of Building Management and Facility Managers

Facility Manager should assign a person who should be responsible of controlling access to the emergency scene and for keeping people away from unsafe areas. Others should be familiar with the locations and functions of controls for building utility, life safety and protection systems. These systems include ventilation, electrical, water and sanitary systems; emergency power supplies; detection, alarm, communication and warning systems; fire suppression systems; pollution control and containment systems; and security and surveillance systems. Personnel should be assigned to operate or supervise these systems as directed by public emergency services if they are on-site.

Site and Facility Plans and Information

Keep a site plan (drawing) handy and available in case of emergencies. Public emergency services have limited knowledge about your facility. And if they act in case of emergency without knowledge of your facility it can be hazardous for the team. Therefore, it is important to document information about your facility. That information is vital to ensure emergency responders can safely stabilize an incident that may occur. Documentation of building systems may also prove valuable when a utility system fails; such as when a water pipe breaks and no one knows how to shut off the water.

Compile a site-plan and plans for each floor of each building. Plans should show the layout of access roads, parking areas, buildings on the property, building entrances, the locations of emergency equipment and the locations of controls for building utility and protection systems. Instructions for operating all systems and equipment should be accessible to emergency responders.

Training and Exercises

It is very important for the personal responsible for Emergency Response team to be trained. They should be familiar with fire detection, alarm, communications, warning and protection systems. Conduct frequent exercise to ensure that the team members are familiar with their role and can carry out assigned responsibilities. Conduct evacuation, sheltering, sheltering-in-place and lockdown drills so employees will recognize the sound used to warn them and they will know what to do. Facilitate exercises to practice the plan, familiarize personnel with the plan and identify any gaps or deficiencies in the plan.

Some simple Steps for Developing the Emergency Response Plan

➢ Periodically review performance objectives of the program.

➢ Review hazard or threat scenarios identified during the risk assessment.

➢ Assess the availability and capabilities of resources for incident stabilization including people, systems and equipment available within your business and from external sources.

➢ Discuss with public emergency services (e.g., fire, police and emergency medical services) to determine their response time to your facility, knowledge of your facility and its hazards and their capabilities to stabilize an emergency at your facility.

➢ Develop protective actions for life safety (evacuation, shelter, shelter-in-place, lockdown).

➢ Develop hazard and threat-specific emergency procedures.

➢ Coordinate emergency planning with public emergency services to stabilize incidents involving the hazards at your facility.

➢ Train personnel so they can fulfill their roles and responsibilities.

➢ Conduct frequent emergency drills to practice your plan.

Preparing facility in event of emergency forecast

As a precautionary measure your focus should be on protection of the building and valuable machinery, equipment and materials inside. Potential damage may be prevented or mitigated by inspecting the following building features, systems and equipment:

➢ Windows and doors

➢ Roof flashing, covering and drainage

➢ Exterior signs

➢ Mechanical equipment, antennas and satellite dishes on rooftops

➢ Outside storage, tanks and equipment

➢ Air intakes

➢ High value machinery

➢ Sensitive electronic equipment including information technology and process controllers

➢ DG backup

➢ Storm water pumps

➢ Cleaning of storm waterlines

➢ Sufficient number of first aid kit

Property conservation activities for specific forecast events include the following:

Winter storm - Keep building entrances and emergency exits clear; ensure there is adequate fuel for heating and emergency power supplies; monitor building heat, doors and windows to prevent localized freezing; monitor snow loading and clear roof drains.

Tropical storms and hurricanes - Stockpile and pre-cut plywood to board up windows and doors (or install hurricane shutters); ensure there is sufficient labor, tools and fasteners available; inspect roof coverings and flashing; clear roof and storm drains; check sump and portable pumps; backup electronic data and vital records off-site; relocate valuable inventory to a protected location away from the path of the storm.

Flooding - Identify the potential for flooding and plan to relocate goods, materials and equipment to a higher floor or higher ground. Clear storm drains and check sump and portable pumps. Raise stock and machinery off the floor. Prepare a plan to use sandbags to prevent water entry from doors and secure floor drains.

References

Means of Egress – U.S. Occupational Safety & Health Administration (OSHA) 29 CFR 1910 Subpart E

    NFPA 101: Life Safety Code® – National Fire Protection Association

    Employee Alarm Systems – OSHA 29 CFR 1910.165

    Evacuation Planning Matrix – OSHA

    Evacuation Plans and Procedures eTool – OSHA

    Design Guidance for Shelters and Safe Rooms – Federal Emergency Management Agency (FEMA 453)

Ergonomics

Study of capabilities and limitations of mental and physical work in different settings. Ergonomics applies anatomical, physiological, and psychological knowledge (called human factors) to work and work environments in order to reduce or eliminate factors that cause pain or discomfort. Ergonomic designs of tools and equipment have helped curtail the occurrence of musculoskeltal disorders and repetitive strain injuries such as carpal tunnel syndrome (CTL).

The major risk in ignoring ergonomics principal will be Musculo Skeletal Disorders. Poor workplace posture is a major cause of back pain, workplace stress and can lead to repetitive strain injuries. This can result in poor employee health's, which ultimately result into reduced productivity, lost time and higher business costs.

Musculoskeletal Ergonomic Injuries are the fastest-growing category (accounting for 1.8 million annual Work Related Musculoskeletal Disorders Up 600% over last 11 years (MSDs).

Symptoms of MSD (Musculo Skeletal Disorders):

> Tightness

> Deformity

> Decreased grip strength

> Loss of function

> Decreased range of motion

- ➢ Numbness

- ➢ Burning

- ➢ Pain/Aching

- ➢ Tingling

- ➢ Cramping

- ➢ Stiffness

Bad postures such as shown below are a primary cause of ergonomic injuries

Propping a phone on shoulder   Slouching over a computer

Common Cause of MSD (Musculo Skeletal Disorders)

- ◆ **Lifting**-continuously lifting and moving

- ◆ **Repetitive motion**(factors: angle, alignment, force, length of time without break)-keying and using the mouse

- ◆ **Contact stress**-hammering

- ◆ **Extreme force**-tightening objects

- ◆ **Vibration**-drilling

- ◆ **Awkward postures**-pulling carts

- ◆ **Awkward sitting postures**-Office works

Lifting Posture

Improper lifting technique is a major cause of back pain, especially among people who must routinely lift heavy objects on the job. This is because most stress from lifting centers in on the lower back. Proper body mechanics shifts this stress from the vulnerable back to the much stronger and more resilient leg muscles.

The object of body mechanics is to maintain balance and control at all times. In addition to preventing injuries, lifting the right way also conserves energy.

The right way to lift:

- ➢ Check the weight of item before lifting.

- ➢ Face the object.

> ➤ Keep your feet apart.

> ➤ Bend your knees and enter a squatting position.

> ➤ Lift the weight slowly. No jerks

> ➤ Keep the object close to your body....

> ➤ Never lift a heavy object above shoulder level.

MSD Risk Control

> ➤ Eliminate unnecessary tasks/movements by redesigning procedures and workstations

> ➤ Take short, frequent breaks

> ➤ Alternate tasks and processes to use different muscle groups

➤ Static positions are your enemy! Whenever you think of it, change position

➤ Small frequent stretches go a long way in preventing MSD's.

Break;

Its very important to take a break in between the work. Frequent breaks and stretch exercise reduces the risk of MSD by many folds.

- Organize tasks around built in breaks

- Eye breaks - blink to moisten eyes every 5-10 minutes. Every 15 minutes or so look away from the screen to distant part of room.

- Micro-breaks - between burst of activity rest the hands, neck and shoulders in a relaxed straight posture.

- Rest breaks - every 60 minutes take a brief 5-minute break and engage in another activity.

- Exercise breaks - every 1-2 hours do gentle stretching exercises

- Instead of using intercom its always good to walk to the desk of person you need to discuss or collect information. It gives you a break and a personnel touch with your colleagues.

Some useful exercises

It is important that you take a break during your work and this break is to be utilized only for some relaxing exercises. These exercises will help in eliminating any chances of MSD.

Neck Exercise

✓ Tilt ear towards shoulder

✓ Reach up and touch top of head with palm to hold in tilted position

✓ Reverse side and repeat

Over Head Reach

◆ Take a deep breathe and reach up overhead with both arms.

◆ Hold 5-10 seconds

◆ Exhale and lower slowly

◆ Repeat 2-3 times

Shoulder Pinch

- Place arms behind head being careful not to press hand into head

- Relax shoulders, and squeeze shoulder blades together while keeping shoulders back and down

- Hold 5-10 seconds.

- Repeat 2-3 times

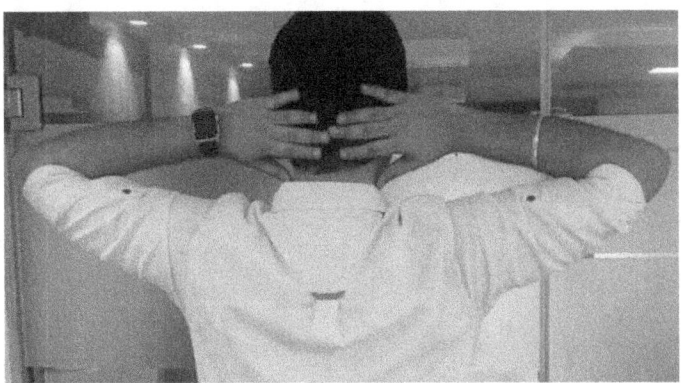

Shoulder Shrug

- Sitting up straight, slowly bring shoulders up toward your ears.

- Hold positions 5-10 seconds

- Then bring the shoulders down and hold

- Repeat 2-3 times

Chair Rotation Stretch

- Sit in chair and place feet flat on floor

- Reach across your body and grab the back of the chair

- Pull gently to increase stretch in mid back

- Hold 5-10 seconds. Repeat 5 times

- Repeat on other side

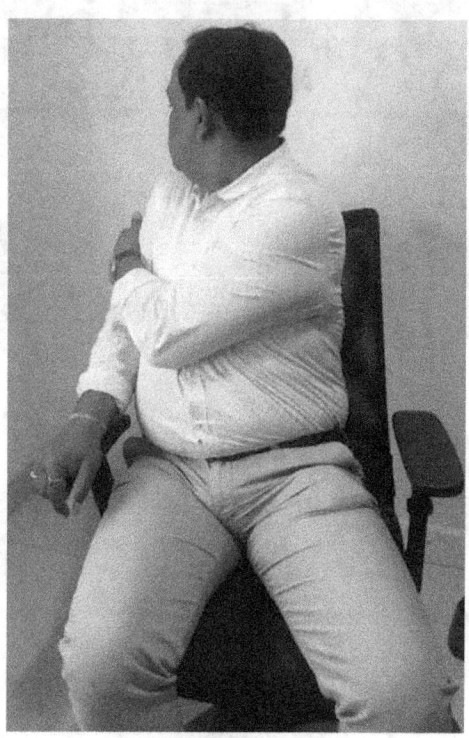

Foot Rotations

- While sitting upright, slowly rotate each foot from the ankles 3 times in one direction

- Then rotate 3 times in the opposite direction

Finger Squeeze

- Squeeze a foam block firmly with all fingers

- Hold for 3 seconds

- Relax your grip

Sitting posture at Workstation

The sitting posture at workstation is very much important. A bad sitting posture will make you uncomfortable and will lead to medical problems. Let check some examples of wrong or right sitting postures;

Unsatisfactory Workstation "Design"

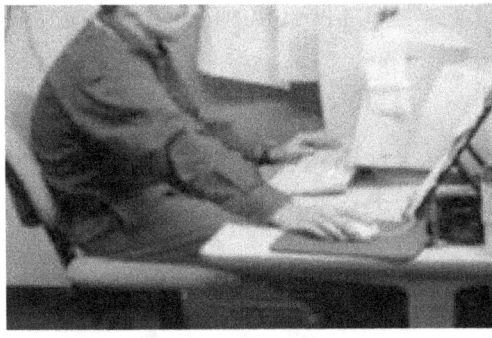

Satisfactory Workstation "Design"

Sitting up straight and Take a break

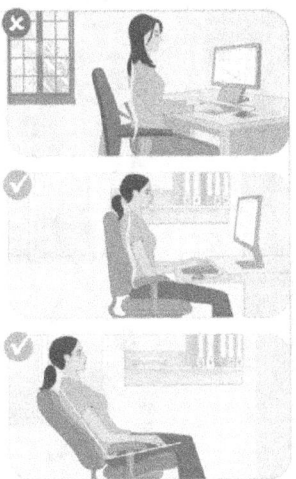

Courtesy: Godrej Furniture

- ➢ Reaching for mouse
- ➢ Monitor too low
- ➢ Keyboard too high, wrists bent
- ➢ Chair too high, feet should be flat on the floor or on a foot rest
- ➢ Document holder too far back
- ➢ No arms on the chair
- ➢ Bad posture, leaning forward

- Use headsets for frequent phone use.
- Monitor at eye level.
- Keep wrists straight, arms close to body, and at a right angle

- Mouse on support next to keyboard.
- Document holder near monitor.
- Foot rest used if feet do not touch the floor.
- Fully adjustable computer task chair.

- One should use ergonomically designed chair, which should be comfortable for you.
- Ensure you follow 90:90 rule and your lower back is always supported by chair back.
- Every 30 minutes lean back in your chair to relax your muscles.

Twelve Important Points for an Ergonomic Work Station

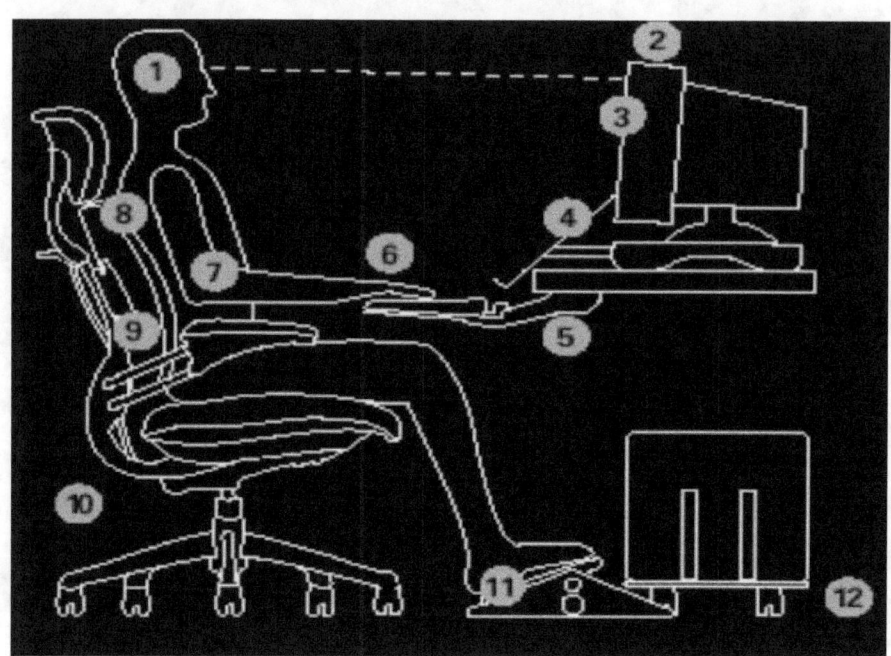

1. Top of monitor at or below eye level
2. Monitor and keyboard centered in front of you
3. No glare in screen
4. Document in line with keyboard & monitor
5. Negative tilt keyboard support
6. Wrist flat & straight
7. Arms & elbow close to body
8. Change posture often
9. Work in a reclined position
10. Take frequent short breaks
11. Feet flat on floor or foot rest
12. CPU off desk

Some Do's & Don'ts

Never be so edgy. This kind of posture will cause strain on your back muscles and ligaments.

Sit straight with back support. If required pull your monitor close or make character large for easy viewing.

Courtesy: Godrej Furniture

FIRE DETECTION & FIGHTING

43% of businesses that suffer a significant fire never reopen.

What is fire?

Fire is the rapid oxidation (chemical reaction) of a material in the exothermic chemical process of combustion, releasing heat, light, and various reaction products.

Fire Tetrahedron: Basic components of a fire are:

- fuel
- heat
- oxygen
- Chain reaction

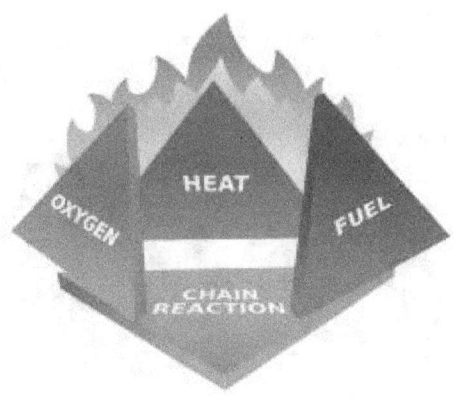

How to extinguish a Fire

Fire can be extinguished by removing any one of the elements of the fire tetrahedron.

1. Starvation: Cut of fuel supply

2. Smothering: Covering the flame completely, which smothers the flame by cutting oxygen supply

3. Cooling: Application of water, which removes heat from the fire faster than the fire can produce it.

4. Stop chain reaction: Stop or interrupt the chain reaction between the fuel, heat and oxygen the fire will be extinguished. Application of a retardant chemical such as Halon/FM 200 to the flame, which retards the chemical reaction itself until the rate of combustion is too slow to maintain the chain reaction.

Class of fire

*   **Class A**

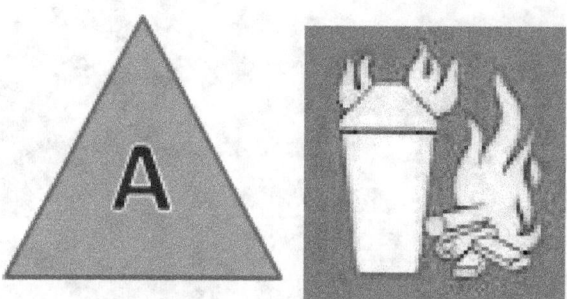

Class "A" fires involve solid materials of an organic nature such as wood, paper, cloth, rubber and plastics that do not melt.

*   **Class B**

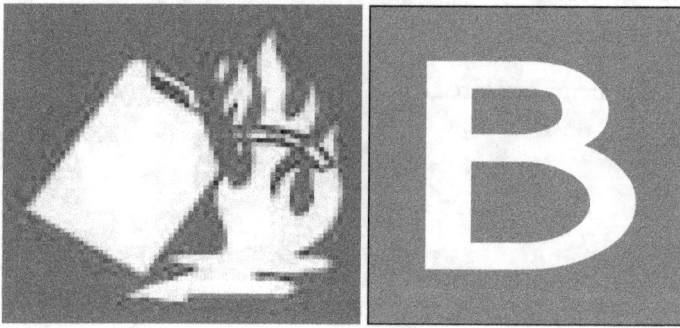

Class B fires involves liquids. They include petrol, diesel, thinners, oils, paints, wax, cooking fat and plastics that melt.

*   **Class C:** Class C fires involve electricity.

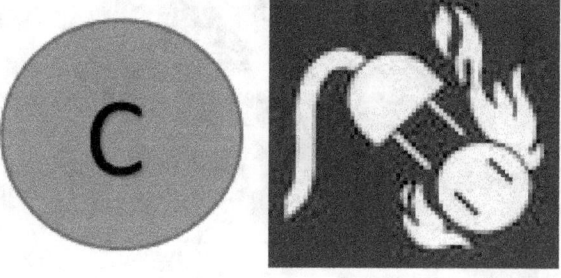

*   **Class D:**

Class D fires involve flammable metals such as magnesium, aluminum, titanium, sodium and potassium.

Fire Chart

CLASS OF FIRE	SUITABLITY ON DIFFERENT CLASSES OF FIRE	EXTINGUISHING EFFECT	EXTINGUISHING AGENT	TYPE OF EXT.
A	WOOD, PAPER AND CLOTH (Furniture, Carpet)	COOLING	WATER	Water
AB	FLAMMABLE LIQUIDS (Petrol, Diesel, Grease, Paint)	BLANKETING	MECHANICAL FOAM	Mechanical Foam
BCD	FLAMMABLE LIQUIDS, GASES, ELCTRICAL INSTALLATIONS like Motors, switch gears and flammable metals.	BLANKETING	DRY POWDER	Dry Chemical Powder
BC	FLAMMABLE LIQUIDS, GASES, COMBUSTIBLE METALS, ELCTRICAL INSTALLATIONS like Motors, Computer Equipment	SMOTHERING	CARBON DIOXIDE	Co2 type

Types of portable fire extinguisher

There are two main types of fire extinguishers:

- **Stored pressure:** In stored pressure units, the expellant is stored in the same chamber as the firefighting agent itself. Depending on the agent used, different propellants are used. With dry chemical extinguishers, nitrogen is typically used; water and foam extinguishers typically use air. Stored pressure fire extinguishers are the most common type.

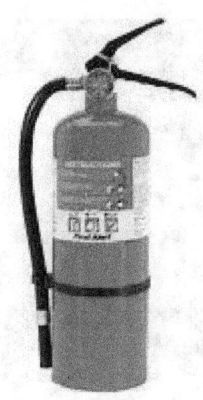

- **Cartridge-operated:** Cartridge-operated extinguishers contain the expellant gas in a separate cartridge that is punctured prior to discharge, exposing the propellant to the extinguishing agent.

  > They have the advantage of simple and prompt recharge

  > Unlike stored pressure types, these extinguishers use compressed CO_2. Cartridge operated extinguishers are available in water, wetting agent, foam, dry chemical (classes ABC and BC), and dry powder (class D) types.

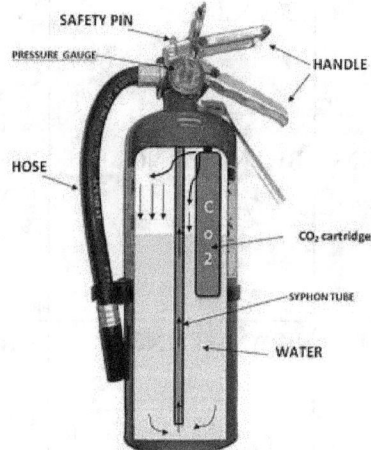

Water type extinguisher

APW (Air Pressurized Water) cools burning material by absorbing heat from burning material. Effective on Class A fires, it has the advantage of being inexpensive, harmless, and relatively easy to clean up.

These extinguishers come in 1.75 and 2.5 gallon units, painted white in the United States and red in India.

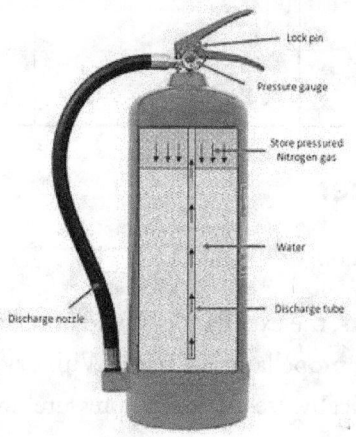

Dry Chemical Powder

Powder based agent extinguishes by separating the four parts of the fire tetrahedron. It prevents the chemical reactions involving heat, fuel, and oxygen and halts the production of fire sustaining "free-radicals", thus extinguishing the fire.

Chemical used: Monoammonium phospate; Sodium- bicrbonate; Potassium- bicrbonate.

Foam

Applied to fuel fires as either an aspirated (mixed & expanded with air in a branch pipe) or non-aspirated form to form a frothy blanket or seal over the fuel, preventing oxygen reaching it.

Unlike powder, foam can be used to progressively extinguish fires without flashback.

Types:

1. AFFF (aqueous film forming foam)

2. AR-AFFF (Alcohol-resistant aqueous film forming foams), used on fuel fires containing alcohol

3. FFF (film forming fluoroprotein) contains naturally occurring proteins from animal by-products

- Clean agents and carbon dioxide:

    Agent displaces oxygen (CO_2 or inert gases), removes heat from the combustion zone (Halotron, FE-36) or inhibits chemical chain reaction (Halons). They are labeled clean agents because they do not leave any residue after discharge which is ideal for sensitive electronics and documents.

Types

1. Halon (including Halon 1211 & Halon 1301), a gaseous agent that inhibits the chemical reaction of the fire. Banned on date.

2. Co2, a clean gaseous agent which displaces oxygen. Not intended for Class A fires, as the high-pressure cloud of gas can scatter burning materials.

Color coding of fire extinguisher

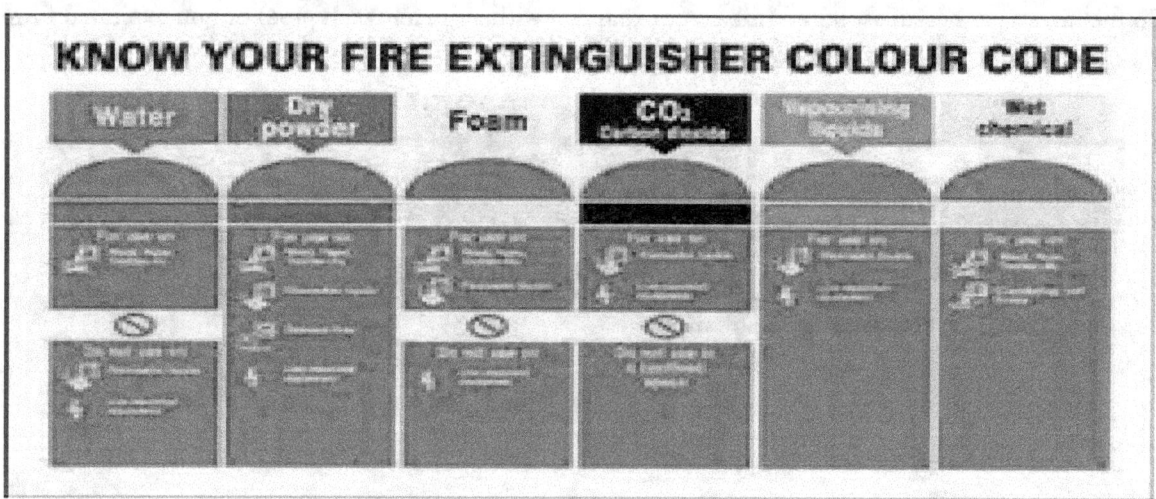

Hydrostatic Test Intervals

Depending on the type(s) of extinguishers you have, they must be emptied and hydrostatically tested at the intervals specified in Table L-1. Extinguisher shells, cylinders, or cartridges that fail a hydrostatic pressure test, or which are not fit for testing, shall be removed from service and the workplace.

Type of Extinguishers	Special Requirements	Test Interval (years)
Foam (soldered brass shells); Soda acid (soldered brass shells)		Must be removed from service
*Soda acid (stainless steel shell); Foam (stainless steel shell)	NOTE: Test self-generating type soda acid and foam extinguishers at 350 psi (2,410 kPa).	5 Years
*Cartridge operated water and/or antifreeze; Stored pressure water and/or antifreeze; Wetting agent; Dry chemical with stainless steel		5 Years

Carbon Dioxide	NOTE: (C02 extinguishers that have a hose assembly equipped with a shut-off nozzle must be tested at 1,250 psi (8,620 kPa). Hose assemblies must also be tested within a protective cage device.	5 Years
Dry chemical, stored pressure, with mild steel, brazed brass or aluminum shells; Dry chemical, cartridge or cylinder operated, with mild steel shells; Dry powder, cartridge or cylinder operated with mild steel shells	NOTE: Dry chemical and dry powder hose assemblies equipped with a shutoff nozzle must be hydrostatically tested at 300 psi (2,070 kPa).	12 Years
Halon 1211; Halon 1301	NOTE: Halon 1211 and all stored pressure extinguishers must be hydrostatically tested at the factory test pressure, not to exceed two times the normal operating pressure.	12 Years

NOTE: All hose assemblies must be hydrostatically tested at the same interval as the extinguisher if it is equipped with a shutoff nozzle at the discharge end.

Hose assemblies passing a hydrostatic test do not require any type of recording or stamping.

Reference: **USA Department of labor website**

- **Fire Planes:**

  Aerial firefighting is the use of aircraft and other aerial resources to combat wildfires. The types of aircraft used include fixed-wing aircraft and helicopters.

  Chemicals used to fight fires may include water, water enhancers such as foams and gels, and specially formulated fire retardants

- **Fire Hydrant System:**

  The system consists of following components:

  ➢ Piping

  ➢ Hydrant Valves

  ➢ Hose reels

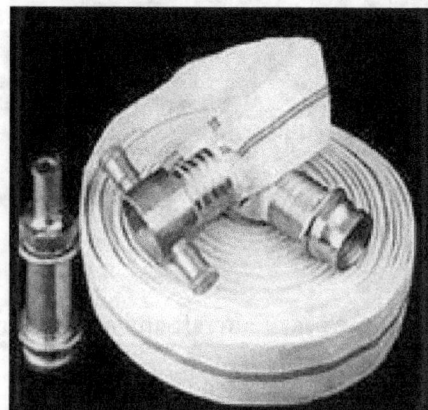

➢ Jockey pump

➢ Hydrant pump

➢ Diesel driven pump

➢ Isolation valves

A fire hydrant is an active fire protection measure. Hydrant system is a mechanism where water is pumped, tapped off and released in the form of a jet through a nozzle to extinguish fire.

1. Fire Escape Hydrant Systems: It's a type of Hydrant system that is extensively in high rise buildings where special outlets are provided on each floor.

2. Yard Hydrant Systems: Yard hydrant system is a method where fire hydrant network is led throughout the designated premises.

Dry-Barrel Fire Hydrants

♦ Used in climates where freezing weather is expected.

♦ Compression, gate, or knuckle-joint opens either with pressure or against pressure.

♦ When hydrant is closed, the barrel is empty from the hydrant down to the main valve.

♦ Any water remaining in hydrant should be drained.

Wet-Barrel Fire Hydrants

♦ Are used in areas that do not have freezing weather.

♦ Are always filled with water to the valves near the discharges.

Fire Hydrant System

♦ Pressure in fire hydrant system is maintained at 8 Kgf/sq. cm.

♦ Jockey pump is designed to maintain pressure in hydrant & sprinkler system. It starts at 4 kgf/sq. cm. and cuts off at 8 kgf/sq. cm.

♦ Diesel driven pump is designed to start when pressure in system drops below 4kg/sq.cm.

♦ Motor driven hydrant pump starts when pressure in system drops below 2 kgf/sq.cm.

♦ Booster pump is started manually in case of underground fire tank gets empty. Booster pump is normally installed on terrace adjacent to overhead water tank.

Fire Sprinkler System

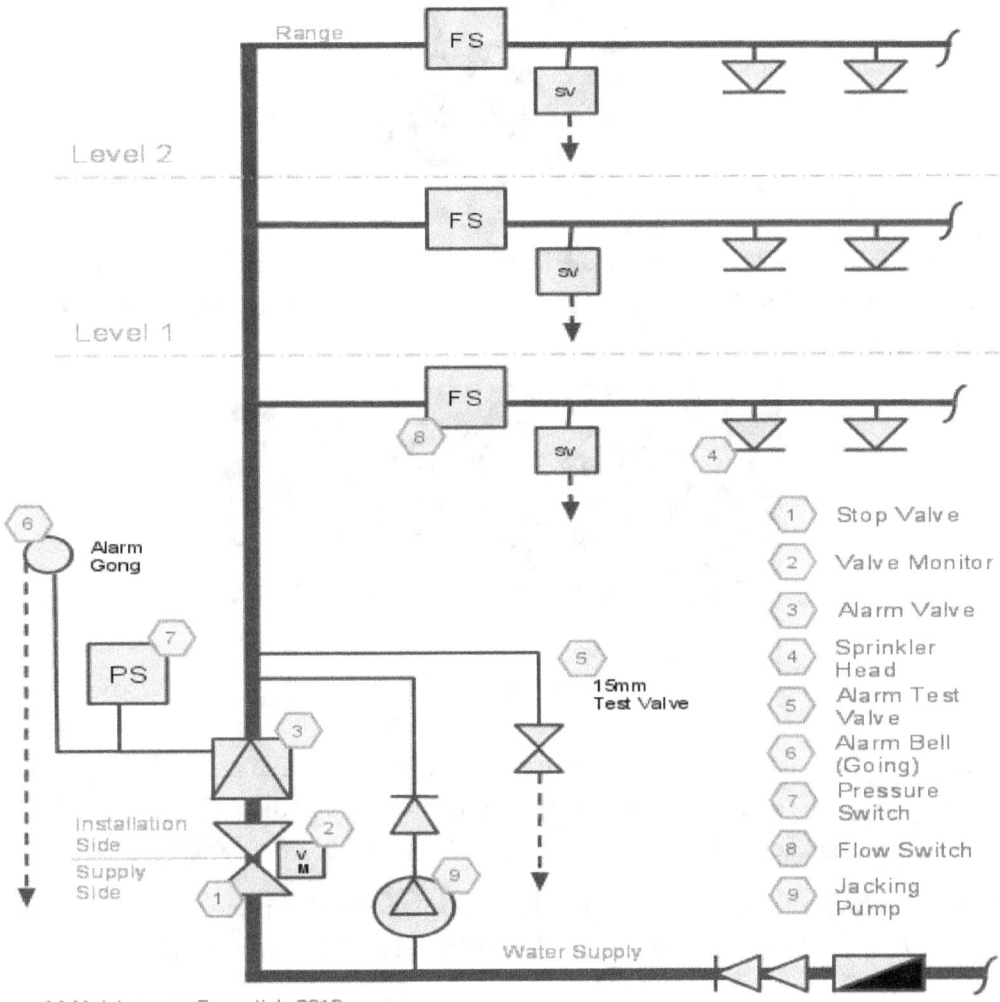

Purpose of each main component

- Stop Valve - The Stop Valve is used to isolate the water supply, and is also called as isolating valve. Painted RED in color.

- Alarm Valve (NRV)- The Alarm Valve is used to control the flow of water into the fire sprinkler system. This is Non Return Valve that is normally closed when the water pressure on the fire sprinkler side of the valve exceeds the water supply pressure.

- Automatic Fire Sprinkler - The Fire Sprinkler is when exposed for a sufficient time to a temperature at or above the temperature rating of the heat sensitive element(glass bulb or fusible link) releases water to flow from only the affected sprinkler.

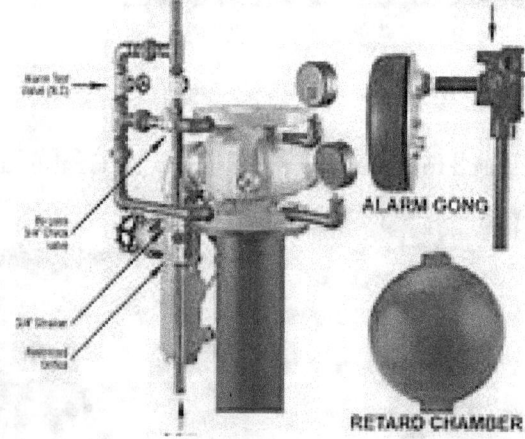

- Alarm Test Valve - The alarm test valve is a small valve, normally secured in the closed position. The alarm test valve is fitted between the sprinkler system side of the alarm valve and the drain. The purpose of the alarm valve is when opened to simulate the flow of water from a single automatic fire sprinkler.

- Motorized Alarm Bell or Gong - The motorized alarm bell or gong is a mechanical device, operated by the flow of water oscillating a hammer that strikes a gong, causing an audible alarm signal.

- Pressure Switch - The pressure switch is an electro-mechanical device that monitors pressure in the system.

- Flow Switch - The flow switch is an electro-mechanical device that monitors the flow of water within an automatic fire sprinkler system.

Flow switches are often fitted with a mechanical delay (up to six minutes) preventing small or minor water flow fluctuations from signaling an alarm.

When sustained water flow is detected by a flow switch, a signal is transmitted to a fire indicator panel. This signal is then used to determine which section (floor) of a fire sprinkler system has water flow.

A fire sprinkler discharges water when the effects of a fire have been detected, such as when a predetermined temperature has been exceeded.

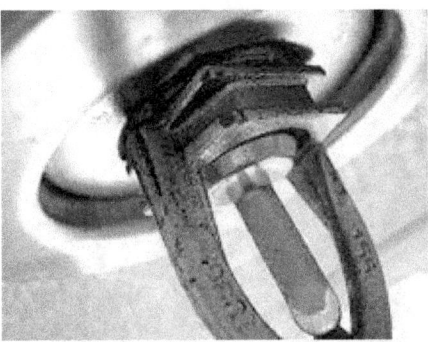

Automatic fire sprinklers utilizes a fusible element, a portion of which melts, or a frangible glass bulb containing liquid which breaks, allowing the plug in the orifice to be pushed out of the orifice by the water pressure in the fire sprinkler piping, resulting in water flow from the orifice. The water stream impacts a deflector, which produces a specific spray pattern.

Frangible bulbs follow a standardized color coding convention indicating their operating temperature.

- The bulb breaks as a result of the thermal expansion of the liquid inside the bulb.

- Under standard testing procedures (135 °C air at a velocity of 2.5 m/s), a 68 °C sprinkler bulb will break within 7 to 33 seconds.

- The sensitivity of a sprinkler can be negatively affected if the thermal element has been painted.

Maximum Ceiling Temperature	Temperature Rating	Temperature Classification	Color Code (with Fusible Link)	Glass Bulb Color
100°F/38°C	135-170°F/57-77°C	Ordinary	Uncolored or Black	Orange (135°F) or Red (155°F)
150°F/66°C	175-225°F/79-107°C	Intermediate	White	Yellow (175°F) or Green (200°F)
225°F/107°C	250-300°F/121-149°C	High	Blue	Blue
300°F/149°C	325-375°F/163-191°C	Extra High	Red	Purple
375°F/191°C	400-475°F/204-246°C	Very Extra High	Green	Black
475°F/246°C	500-575°F/260-302°C	Ultra High	Orange	Black
625°F/329°C	650°F/343°C	Ultra High	Orange	Black

Sprinkler Bulbs

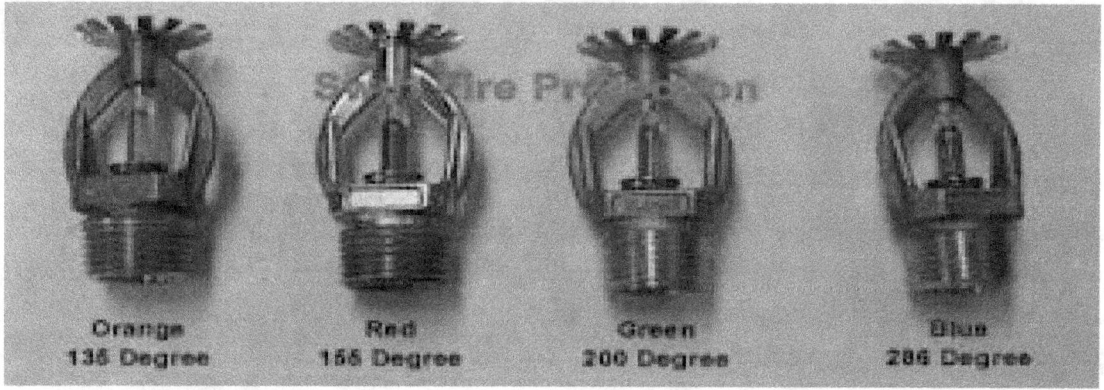

Fire Sprinkler heads are color coded. The temperature mentioned is in Fahrenheit.

Installation of Sprinkler

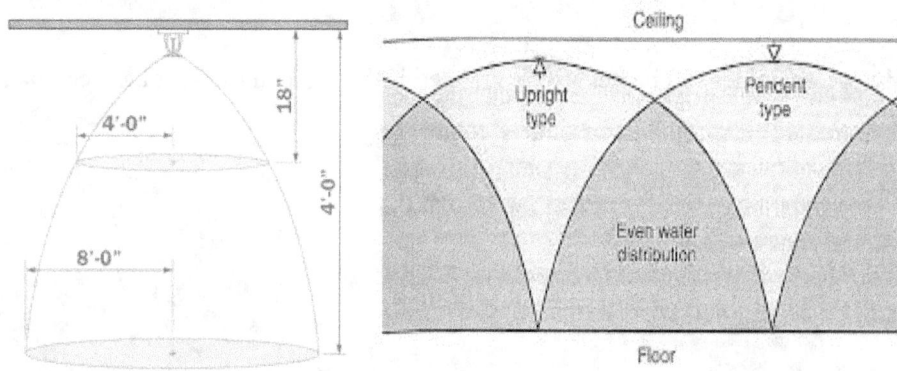

The gap between two sprinklers as per NFPA13 shall be as appended below (Please note that the appended table is only for reference and shouldn't be used for designing. Designing has to be done by a qualified engineer)

Area	Sq. ft. area per head	Distance
Office, Educational, Religious, Institutional, Hospitals, Restaurants, Clubs, Theaters, etc.	130-200 SF per head	6ft to 15 ft.
Mills, Manufacturing, Processing, Machine Shops, Repair Garages, Post Offices, Bakeries, Wood Machining and Assembly, Auto Parking, etc.	130 SF per head	15 ft.
Plastic Processing, Chemical Spraying, Metal Extruding, Printing, Varnishing, Painting, etc.	90-130 SF per head	12 ft.

> Distance from Ceiling: minimum 1 inch to 12 inch

> for unobstructed construction. The minimum 1" is typical; however, concealed, recessed, and flush sprinklers may be mounted less than 1" from the ceiling and shall be installed based on their listing.

Maximum Distance from Wall: half (1/2) of the maximum distance between sprinkler heads.

Gas Suppression System

- The system uses inert gases and chemical agents to extinguish a fire.

- This system is basically used in critical areas such as Laboratory, Server rooms, Data Centers and Ammunition magazine.

- The system uses argon gas for use in extinguishing fire, without damaging equipment.

- The bright green shoulder of the H cylinders signals an inert gas.

- Red is used here for argon, but other colors are used for argon in various systems, which are not fully standardized across the world.

Gases used

- Reduction of heat: HFC-227ea (MH227, FM-200), Novec 1230, HFC-125 (ECARO-25), FS 49 C2.

- Reduction or isolation of oxygen: Argonite/ IG-55 (ProInert), CO2 , IG-541 Inergen, and IG-100 (NN100).

- Inhibiting the chain reaction of the above components: FE-13, FE-227, FE-25, MH227, FM-200, Halons, Halon 1301, Freon 13T1, NAF P-IV, NAF S-III, and Triodide (Trifluoroiodomethane).

- Halon (UNEP banned it in the Montreal Protocol treaty in 1987 due to the Ozone depleting effect of Halon gases.)

- FS 49 C2 acts similar to as an inert gas. The heat produced from the combustion process is absorbed by the extinguishant.

- The difference from inert gases and FS 49 C2 is that it takes less gas to suppress a fire, and therefore less space is consumed for storage of gas to a fire extinguishing system. How much space saved will depend on the pressure used when filling the cylinder. It may vary between a 50-90% saving of storage space.

Safety Precaution

The Gas releases in enclosed spaces present a risk of suffocation. To prevent deaths due to suffocation, additional life safety systems are typically installed with a warning alarm that precedes the agent release.

The warning, usually an audible and visible alert, advises the immediate evacuation of the enclosed space.

Inert Gas Fire Suppression

- Concept Fire Suppression's IG55 Inert Gas Fire Suppression Agent is a standard 50:50 mix of two inert gases Nitrogen and Argon.

- There are 4 types of agent generally available namely:-

Designated Name	Trade Name	Composition
IG-01	Argotec, Argon fire	Argon 100%
IG-55	Argonite	Argon 50%, Nitrogen 50%
IG541	Inergen	Argon 40%, Nitrogen 52%, CO2 8%

- Nitrogen and Argon gases occur naturally in the atmosphere and are therefore the ultimate in an environmentally safe fire suppression product that is also safe for people. They have a Global Warming Potential and Ozone Depleting Potential of 0 and do not decompose on application to a flaming fire.

- IG-01 and IG100 are only occasionally used but the standard mixes of Argonite and Inergen are standard products, with most manufactures carrying stocks.

Major Foam Fire Fighting System

The system is used to fight oil fire. It basically uses fire hydrant system.

The system is consisting of:

> Branch pipe

> Intermediate pipe with suction port

> ➤ Portable tank or Can

> ➤ Canvas hose pipe

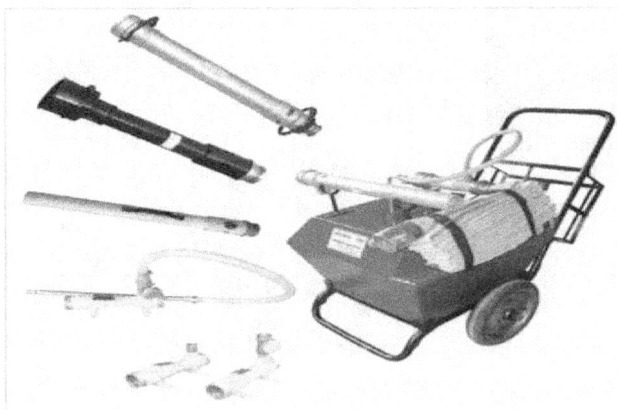

FOAM: A firefighting foam is simply a stable mass of small air-filled bubbles, which have a lower density than oil, gasoline or water.

- Foam is made up of three ingredients - water, foam concentrate and air.

- When mixed in the correct proportions, these three ingredients form a homogeneous foam blanket.

HOW FOAM EXTINGUISHES A FLAMMABLE LIQUID FIRE

Fire burns because of four elements. Under normal circumstances if any one of the elements is removed/interfered with, the fire is extinguished. Firefighting foam does not interfere in the chemical reaction.

Foam works in the following ways:

- The foam blankets the fuel surface smothering the fire.

- The foam blanket separates the flames/ignition source from the fuel surface.

- The foam cools the fuel and any adjacent metal surfaces.

- The foam blanket suppresses the release of flammable vapors that can mix with air.

Mechanical foam concentrates that are the most common types currently used by fire fighters:

- Aqueous Film Forming Foam (AFFF)

- Alcohol Resistant (AR-AFFF)

- Synthetic – medium or high expansion types (detergent)

- Class "A" Foam Concentrate

- Wetting Agent

- Fluoroprotein

- Protein

- Film Forming Fluoroprotein (FFFP)

Fire Alarm System

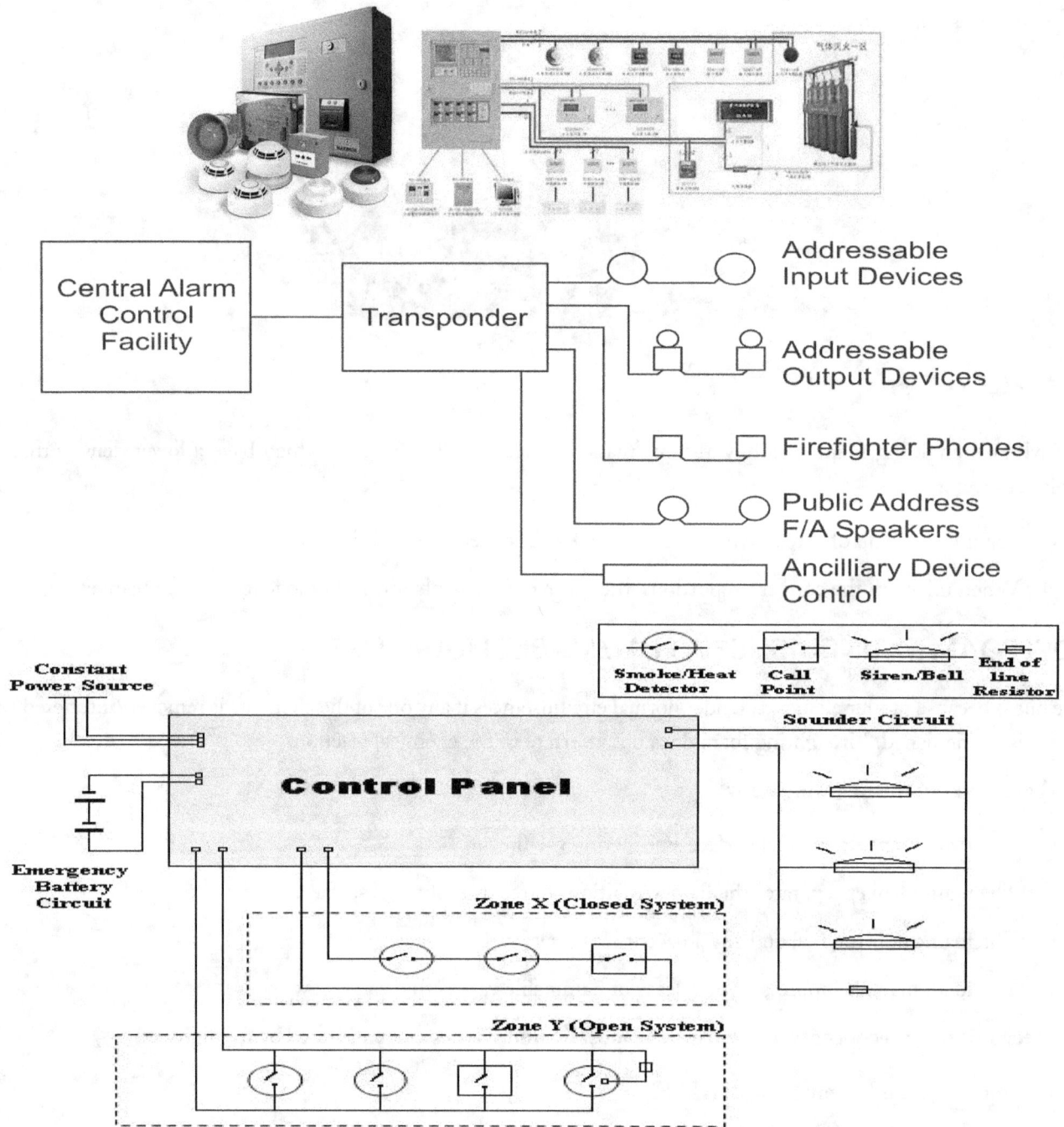

Types of Smoke Detectors

There are two common types of smoke detectors.

Each type is distinguished by its detection method.

- The most common smoke detector uses ionization sensors to detect smoke.

- The other type of smoke detector uses photoelectric sensor to detect smoke.

Ionization Detectors

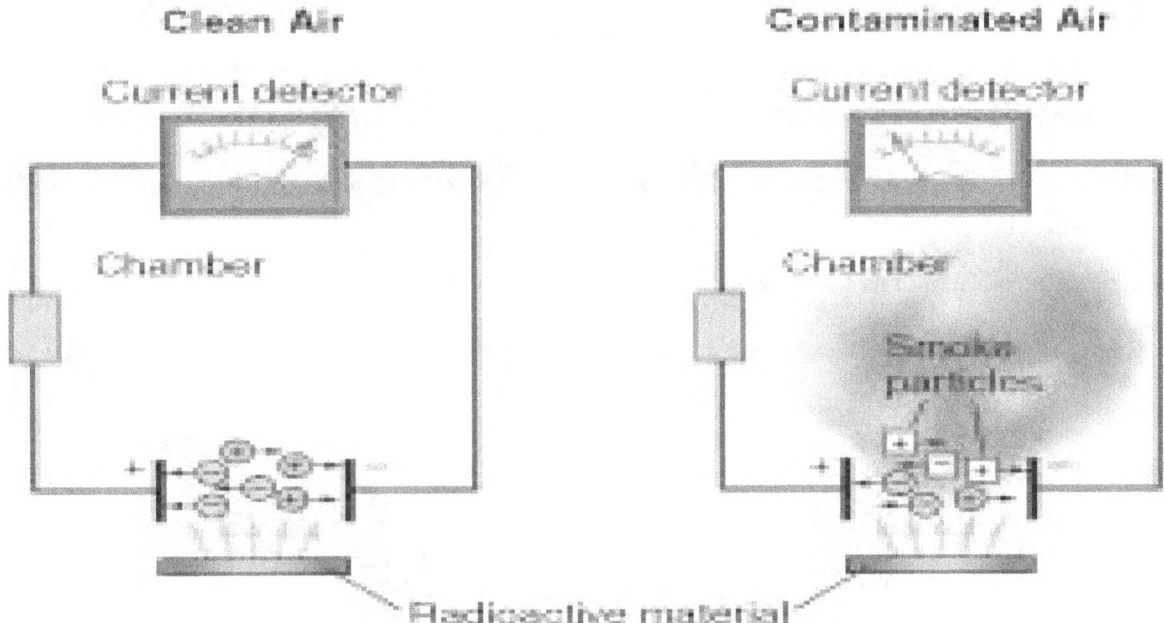

Clean Air	Contaminated Air
The radioactive material releases charged particles into the chamber, and a small electric current flows between two plates.	When smoke particles enter the chamber, they neutralize the charged particles and interrupt the current flow. The detector senses the interruption and activates the alarm.

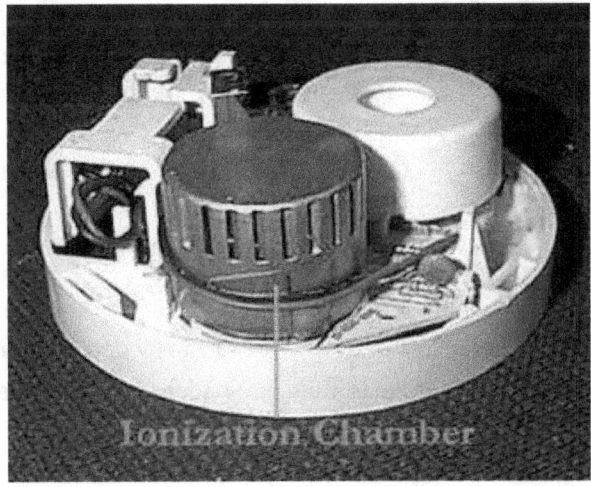

Ionization Chamber

Ionization Detectors

- Ionization detectors detect the presence of invisible particles (less than .01 micron in size) in the air.

- Inside the detector, there is a small ionization chamber that contains an extremely small quantity of radioactive isotope called Americium-241.

Photoelectric Detectors

Photoelectric detectors detect the presence of visible particles (larger than 3 microns) in the air.

Inside the detector, there is a light emitting diode (LED) that directs a narrow beam of infrared light across the detection chamber. When smoke or particles enter the chamber, the infrared light beam is scattered.

A photodiode or photodetector, usually placed 90 degrees to the beam, will sense the scattered infrared light and when a preset amount of light is detected, the alarm will sound.

Photoelectric detectors are not as sensitive and are designed to detect cool or slow-moving (smoldering) fires that produce a lot of smoke.

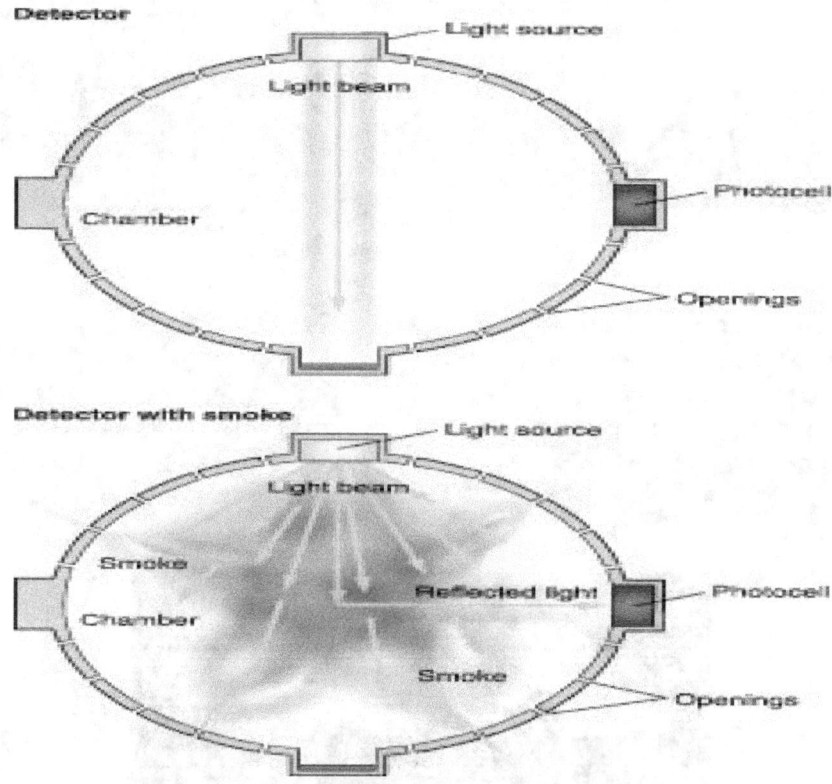

Heat Detectors

Heat detectors are effective to warn about fire hazards at the early stage by detecting the increased heat around the area. With easy installation in both indoor & outdoor areas, these are widely utilized in various residential and commercial establishments.

Beam Detector

Beam Detectors are cost effective when compared with using point detection and suitable for heights way above the operating limits of conventional smoke detection.

It can be used in cinema halls, auditorium and storage area.

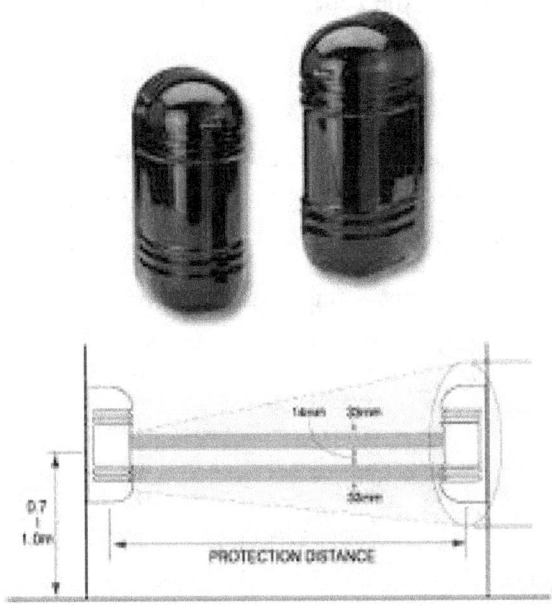

The operational range of a standard beam detector is 100M with coverage up to 15m Wide. Immediately we can see that covering 1500m2 with one pair of beam detectors is quite impressive. This also means that the beam must be connected to its own zone on a conventional fire alarm panel or its own software zone on an analogue addressable control panel. In the case of the latter a short circuit isolator will also be required as no zone can exceed 2000m2 under BS5839 part One:1988.

Manual Fire Alarm

- Fire alarm pull station: A fire alarm pull station is an active fire protection device, usually wall-mounted. When activated, initiates an alarm on a fire alarm system. In its simplest form, the user activates the alarm by pulling the handle down, which completes a circuit and locks the handle in the activated position, sending an alarm to the fire alarm control panel.

Types of Fire Alarm Panel

1. **Conventional Fire Alarm Panel:**

   A conventional Fire Alarm Control Panel employs one or more circuits, connected to sensors wired in parallel.

   A conventional or non-addressable system allows multiple devices to be connected to a single zone. However, when an alarm sounds, you won't know where in the zone that fire is located. These systems are used in older or smaller buildings.

2. Addressable fire alarm panel:

   This system will identify the specific "address" or location of an alarm. Addressable systems are usually installed in larger buildings because it saves time in identifying the exact location of a fire. The main panel will identify the device and its exact location which will help you to know if the alarm is responding to heat, smoke or fire.

   Addressable Fire Alarm Control Panels employ one or more Signaling Line Circuits, slang - usually referred to as loops or SLC loops – ranging between one and thirty.

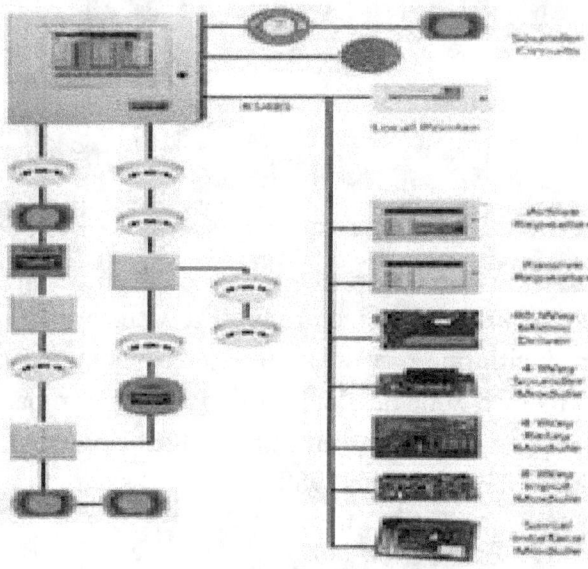

Vesda System

♦ An aspirating smoke detector (ASD), consists of a central detection unit which draws air through a network of pipes to detect smoke. The sampling chamber is based on anephelometer that is capable of detecting the presence of smoke particles suspended in air by detecting the light scattered by them in the chamber.

- In most cases aspirating smoke detectors require a fan unit to draw in a representative sample of air from the protected area through its network of pipes, such as is the case for Wagner, Safe Fire Detection's ProSeries and Xtralis ASD systems.

- Aspirating smoke detectors are highly sensitive, and can detect smoke before it is even visible to the human eye.

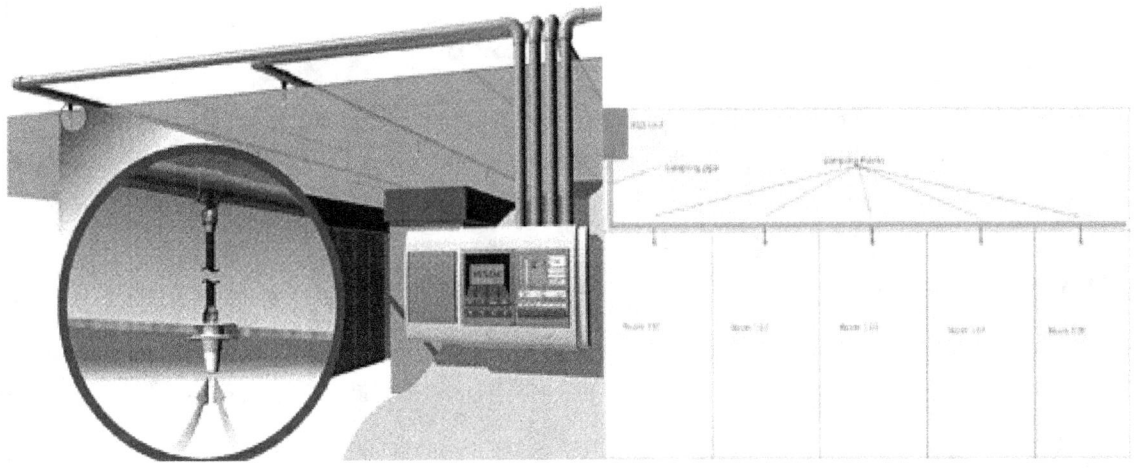

Accessories for Fire Fighting

Breathing Apparatus

Whenever there is a fire in a building, the fire tend to generate lots of smoke. These smoke results in choking and causes death for a trapped person. Maximum deaths due to fire are caused by smoke choking. Moreover, in such situation even a rescue mission can't be accomplished. The fire fighters or the rescue team should always use breathing apparatus whenever they are entering any building on fire. This breathing apparatus is also known as SCBA (Self Containing Breathing Apparatus). Every commercial and residential building should have one set of breathing apparatus.

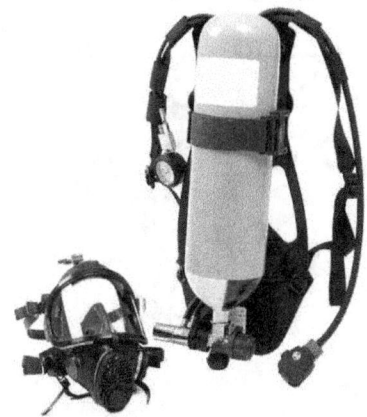

Fire Axe

Firemen's axe is an important tool used by fire fighters. The axe is used to break through a locked door for rescue and firefighting. Striking tools can be used to strike a surface, such as chopping a roof for ventilation purposes, or to strike another tool, such as an axe striking a Halligan tool during forcible entry.

This axe is also used to chop of electrical line to cut of power supply to the building in fire. The handle of this axe is electricity resistant up to 2KV. The fireman axe is a pick head axe which has a pick shaped pointed poll and a chrome plated blade. Its handle has required grip while breaking down doors and windows in case of an emergency evacuation.

Aluminum Proximity Suit

This suit is made of vacuum-deposited aluminized materials. This material makes the suit fire resistant and is majorly used by firefighters to enter a building on fire. There are three basic types of aluminized suits. There are three types of fire proximity suit.

- A. Approach suit used for work in the general area of high temperatures such as steel mills and smelting facilities. Maximum ambient heat protection is about 200°F (93 °C).

- B. Proximity suit used for ARFF. Maximum ambient heat protection is about 500°F (260 °C).

- C. Entry suit used for entry into extreme heat and situations requiring protection from total flame engulfment. Most commonly made of Zetex or Vermiculite and not aluminized. Maximum ambient heat protection is about 2,000°F (1,093 °C)) for short duration and prolonged radiant heat up to 1,500°F (816 °C)

For complete protection while firefighting and entry in building in fire ARFF requires aluminized hood or helmet cover with neck shroud, aluminized jacket and pants with vapor-barrier insulated liner, aluminized lined gloves, aluminized ARFF boots, and aluminized covers for SCBA bottles or suits that cover the air pack.

There are three primary materials used in proximity ensembles: aluminized glass, Nomex and Rayon.

Proximity Suit

Proximity suits are designed for exterior rescue operations. Modern Apparels' fire proximity suit is made of 16-ounce imported aluminized glass fiber fabric with dual mirror having 90 percent reflection of heat as an outer layer and with a woolen fabric lining. An additional vapor-barrier layer is also used for higher temperature.

Fire Entry Suit

This suit is used to enter in to a premises on fire. It weighs approx. 48 pounds and can protect against heat up to 1650 deg C.

Firefighters Helmet

This helmet protects a firefighter from heat, cinders and falling objects. These helmets are made from composite fire resistant material and are lightweight.

National Fire Protection Association

NFPA, established in 1896, is an nonprofit organization who prepares & advocates consensus codes & standards, research, training & education

Fire Fighting Shafts

As per UK Building Regulations, buildings with a floor at more than 18m above fire service vehicle access level, or with a basement at more than 10m below fire service vehicle access level, should be provided with firefighting shafts containing firefighting lifts, firefighting stairs and firefighting lobbies which are combined in a protected shaft known as the firefighting shafts.

Again, buildings with two or more basement stories each exceeding 900m2 in area should be provided with firefighting shafts, which need not include fire-fighting lifts.

Currently, several classes of buildings (industrial, storage, and commercial buildings) less than 18m in height are also required to have a fire-fighting shaft.

Safety Standards

As per UK Building Regulations2000 Approved Document B, FIRE SAFETY, the width of escape stairs has been regulated as below:

i. Stairs with a rise of more than 30m should not be wider than 1400mm unless provided with a central hand rail IITK-GSDMA-Fire01-V5.0 8 Review of Fire Codes

ii. Stairs wider than 1800mm should be provided with a central handrail.

Evacuation Strategies

◆ Ensuring life safety is the most essential aspect of Building Codes. HighIITK-GSDMA-Fire01-V5.0 9. NFPA 72 and Australian Standard AS2200.

◆ In high-rise buildings with large number of occupants it has been found that single phase evacuation is a time consuming process and is impracticable. This has led to a system of evacuation known as phased evacuation in which the building is evacuated in different phases in the event of fire.

Evacuation by Fire Lift

◆ BS 5588: Fire Prevention in the Design, Construction and Use of Buildings, Part 5: Code of Practice for Fire Fighting Stairs and Lifts recommends the use of lifts under emergency situations with provision for protection against fire, smoke and heat.

National Building Code

RENEWAL OF FIRE CLEARANCE

On the basis of undertaking given by the Fire Consultant/Architect, the Chief Fire Officer shall renew the fire clearance in respect of the following buildings on annual basis:-

- Public entertainment and assembly

- Hospitals

- Hotels

- Underground shopping complex

    ➢ As per NBC 4.8.1: The width of the main street on which the building abuts shall not be less than 12.0 m.

    ➢ If there are any bends or curves in the approach road, sufficient width shall be permitted at the curve to enable the fire tenders to turn, the turning circle shall be at least of 9.0 m. radius.

    ➢ The approach to the building and open spaces on its all sides should be(see Building Bye-Laws 4.8 and 4.9) up to 6.0 m. same shall be of hard surface capable of taking the weight of fire tender, weighing up to 22 tones for low rise building and 45 tones for building 15 m., and above in height. The said open space shall be kept free of obstructions and shall be motor able.

Fire Exit

➢ Exits shall be so arranged that they may be reached without passing through another occupied unit, except in the case of residential buildings.

➢ Firefighting equipment where provided along exits shall be suitably located and clearly marked but must not obstruct the exit way and there should be clear indication about its location from either side of the exit way.

As per NBC 4.8.3: Arrangement of Exits

◆ Exits shall be so located so that the travel distance on the floor shall not exceed 22.50 m. for residential, educational, institutional and hazardous occupancies and 30.0 m. for assembly, business, mercantile, industrial and storage occupancies.

◆ Whenever more than one exit is required for a floor of a building they shall be placed as remote from each other as possible. All the exits shall be accessible from the entire floor area at all floor levels.

◆ The travel distance to an exit from the remote point shall not exceed half the distance as stated above except in the case of institutional occupancy in which case it shall not exceed 6.0 m.

As per NBC 4.8.5: Staircase Requirement

◆ For buildings identified in Bye-Laws No. 1.13 VI (a) to (m), there shall be a minimum of two staircases and one of them shall be enclosed stairway and the other shall be on the external walls of building and shall open directly to the exterior, interior open space or to any open place of safety.

◆ Single staircase may be accepted for educational, business or group housing society where floor area does not exceed 300 sq. m. and height of the building does not exceed 24 m.

As per NBC 4.8.5: Minimum Width Provisions for Stairways

The following minimum width provisions shall be made for each stairway

a) i) Residential low rise building   0.9 m.

    ii) Other residential building e.g. flats, hostels, group housing, guest houses, etc   1.25 m.

b) Assembly buildings like Auditorium, theatres and cinemas   2.0 m.

c) All other buildings including hotels   1.5 m.

d) Institutional building like hospitals   2.0 m.

e) Educational building like School, Colleges.   1.5 m.

Doorways

◆ Every doorway shall open into an enclosed stairway, a horizontal exit, on a corridor or passageway providing continuous and protected means of egress.

◆ No exit doorways shall be less than 100 cm in width and 150 cm in case of hospital and ward block. Doorways shall not be less than 200 cm in height.

◆ Exit doorways shall open outwards, that is away front the room but shall not obstruct the travel along any exit.

◆ No door when opened shall reduce the required width of stairway or landing to less than 100 cm. Overhead or sliding door shall not be installed.

4.9.4 Provision of exterior Open Spaces around the Building

Sl. No.	Height of the Building Up to (m.)	Exterior open spaces to be left out on all sides in m. (front rear and sides in each plot)
1	10	As per prescribed set backs
2	15	5
3	18	6
4	21	7
5	24	8
6	27	9
7	30	10

(*Contd.*)

Sl. No.	Height of the Building Up to (m.)	Exterior open spaces to be left out on all sides in m. (front rear and sides in each plot)
8	35	11
9	40	12
10	45	13
11	50	14
12	55 and above	16

NBC 7.14 ELECTRICAL SERVICES

Electrical Services shall conform to the following:

a) The electric distribution cables/wiring shall be laid in a separate duct shall be sealed at every floor with non-combustible material having the same fire resistance as that of the duct. Low and medium voltage wiring running in shaft and in false ceiling shall run in separate conduits.

b) Water mains, telephone wires, inter-com lines, gas pipes or any other service lines shall not be laid in ducts for electric cables.

c) Separate conduits for water pumps, lifts, staircases and corridor lighting and blowers for pressuring system shall be directly from the main switch panel and these circuits shall be laid in separate conduit pipes, so that fire in one circuit will not affect the others. Master switches controlling essential service circuits shall be clearly labeled.

d) The inspection panel doors and any other opening in the shaft shall be provided with airtight fire doors having fire resistance of not less than 1 hour.

e) Medium and low voltage wiring running in shafts, and within false ceiling shall run in metal conduits. Any 230 voltage wiring for lighting or other services, above false ceiling should have 660V grade insulation. The false ceiling including all fixtures used for its suspension shall be of noncombustible material.

f) MCB and ELCB shall be provided for electrical circuit.

NBC 7.20.1 First Aid /Fixed Fire Fighting /Fire Detection Systems and other Facilities

Provision of fire safety arrangement for different occupancy from. SI no. 1 to 23 as indicated below shall be as per Annexure 'A' 'B' & 'C'.

1. Access

2. Wet Riser

3. Down Comer

4. Hose Reel

5. Automatic Sprinkler System

6. Yard Hydrant

7. U.G. Tank with Draw off Connection

8. Terrace Tanks

9. Fire Pump

10. Terrace Pump

11. First Aid Fire Fighting Appliances

12. Auto Detection System

13. Manual operated Electrical Fire Alarm System

14. P.A System with talk back facility

15. Emergency Light

16. Auto D.G. Set

17. Illuminated Exit Sign

18. Means of Escape

19. Compartimentation

20. MCB /ELCB

21. Fire Man Switch in Lift

22. Hose Boxes with Delivery Hoses and Branch

23. Refuge Area

NBC 7.21 STATIC WATER STORAGE TANK

◆ A satisfactory supply of water exclusively for the purpose of firefighting shall always be available in the form of underground static storage tank with capacity specified in Annexure-A with arrangements of replenishment by town's main or alternative source of supply @ 1000 liters per minute.

◆ The static storage water supply required for the above mentioned purpose should entirely be accessible to the fire tenders of the local fire service.

◆ To prevent stagnation of water in the static water tank the suction tank of the domestic water supply shall be fed only through an over flow arrangement to maintain the level therein at the minimum specified capacity.

◆ The static water storage tank shall be provided with a fire brigade collecting branching with 4 Nos. 63mm dia instantaneous male inlets arranged in a valve box with a suitable fixed pipe not less than 15 cm dia to discharge water into the tank. This arrangement is not required where down comer is provided.

NBC 7.22 AUTOMATIC SPRINKLERS

Automatic sprinkler system shall be installed in the following buildings:

a) All buildings of 24 m. and above in height, except group housing and 45 m. and above in case of apartment / group housing society building.

b) Hotels below 15 m. in height and above 1000 sq. m. built up area at each floor and or if basement is existing.

c) All hotels, mercantile, and institutional buildings of 15 m. and above.

d) Mercantile building having basement more than one floor but below 15 m. (floor area not exceeding 750 sq. m.)

e) Underground Shopping Complex.

f)   Underground car/scooter parking /enclosed car parking.

g)   Basement area 200 sq. m. and above.

h)   Any special hazards where the Chief Fire Officer considers it necessary.

i)   For buildings up to 24 m. in height where automatic sprinkler system is not mandatory as per these Bye-Laws, if provided with sprinkler installation following relaxation may be considered.

    i.   Automatic heat/smoke detection system and M.C.P. need not be insisted upon.

    ii.   The number of Fire Extinguisher required shall be reduced by half.

7.24 FIRE ALARM SYSTEM

All buildings of 15 m. and above in height shall be equipped with fire alarm system, and also residential buildings (Dwelling House, Boarding House and Hostels) above 24 m. height.

NBC 2005 Part IV' TAC' NFC

The National Building Code 2005 Part IV'TAC' NFC classifies the Buildings into various hazard categories depending upon the type and usage of the building.

Accordingly' the fire hydrant and Automatic sprinkler system is detailed in NBC as per the Building requirement.

The system comprises of;

- Electric Driven/diesel Engine driven single/ Multistage horizontal or vertical High pressure Pumps to feed Hydrant System & Sprinkler system.

- Pressurization/Jockey Pump to maintain the minimum pressure required as per the system Design.

- Electrical control Panel for automatic manual control.

- 63mm Fire hydrants (Gun metal/SS)

- 20mm First Aid Hose Reels

- Branch Pipes (gun metal/SS)

- 63mm RRL/CP Hose Pipes

- Water monitors

- Flow Switches' Pressure switches

- Sprinklers (different types to suit the usage)' spray nozzles

- System control accessories as strainers' check valves' sluice valves' Butterfly valves' Orifice plates' Air Vessel' Air release Valve' Ball Valves' Water Gong Bell' etc.

Maintenance of Fire Extinguishers

The maintenance of all kinds of portable fire extinguishers is to be undertaken as per IS 2190: 2010 Standard.

Clause: 11.1, The owner, or agent, or occupant of the property is responsible for inspection, maintenance & recharging of the extinguisher.

Clause 11.5: Periodic inspection of fire extinguisher should include:

i.   No obstruction to access or visibility

ii.   Operating instructions on nameplate legible and facing outward

iii. Safety seals and tamper indicators are not broken or missing

iv.   Fullness determined by weighing or lifting

v.   Examination for physical damage, corrosion, leakage, or clogged nozzle

vi.   Pressure gauge reading or indicator in operable range or position

vii. Condition of tiers, wheels, carriage, hose of wheel mounted fire extinguisher

Standard: IS 2190: 2010, Clause 11.10.03.02: Typical condition indicating that an extinguisher is unsafe for use. Potentially most serious hazard of extinguisher is the sudden uncontrolled release of pressure or ejection of part, which could be caused due to any one of below cause:

i.   Corrosion, wear & tear or damage to the threads of any pressure retaining parts.

ii.   Corrosion of welds and

iii. Extensive general corrosion or severe pitting

Quarterly Maintenance of Fire Extinguisher, clause 11.13

- Clean the extinguisher

- Check the nozzle outlet & vent holes as well as threaded portion of the cap for clogging & check the plunger is free & moving.

- Ensure cap washer is intact and grease the cap thread & plunger.

- Make sure the extinguisher is in proper condition and is not accidently discharged. In case of pressure extinguisher, pressure gauge to be checked for correct pressure.

- Check component of extinguisher

Annual Maintenance

- Discharge extinguisher if due for discharge test

- Dismantle & check the extinguisher internally for any kind of rust or corrosion.

- If there are visible sign of rust, then wash cylinder thoroughly and fill it with clean water for 24 hour and observe the surface again. If you still find rust marks, it means the plating thickness is not sufficient and it means the surface needs plating again or phospated.

- IN case of Co2 or Clean agent extinguisher if recharging is not due check the weight and pressure, if it is within limit then the content need not be discharged.

- Inspect the cylinder & its component for any kind of physical damage.

Refilling & Testing of Extinguishers

- Two years: a. Water type extinguisher b. Mechanical foam type, 135 liter fire engine foam type

- Three years: BC & ABC Powder extinguisher confirming to IS 4308, IS 14609
- Five years:
  i.   Portable fire extinguisher water type 9 ltr. 50 ltr (cartridge)
  ii. Portable fire extinguisher mechanical foam type 9 lte, 50 ltr (cartridge)
  iii. Co2 type extinguisher portable & trolley mounted.
  iv. DCP extinguisher high capacity trolley mounted.
  v.   DCP for metal fire.
  vi. Clean agent fire extinguisher

Annexure E, Clause 12.2.1 & 12.2.2 (g)

- E1. Every extinguisher should be hydraulically pressure tested as per the appended schedule. If any leakage or distortion found during pressure testing then the extinguisher need to be replaced.
- E2. The C02 type extinguisher should be pressure tested every time it is sent for recharging (after periodic discharge or otherwise) to the pressure specified in the relevant Indian standard specifications.

Refilling & Operational Test Schedule-Fire Extinguisher

- D-1.1 Once in two years:
  a) Portable fire extinguisher, water type, stored pressure
  b) Portable fire extinguisher, mechanical foam type, stored pressure
  c) 135 liter fire engine, foam type
- D-1.2 Once in three years:

BC & ABC type fire extinguisher confirming to IS4308 & IS14609 respectively

- D-1.3 Once in five years:
  a) Portable fire extinguisher, water type, 9 ltr (gas cartridge)
  b) Portable fire extinguisher, mechanical foam type, 9 ltr (cartridge type)
- D-1.3 Once in five years:
  c) Portable fire extinguisher, water type, 50 ltr (gas cartridge)
  d) Portable fire extinguisher, mechanical foam type, 50 ltr (cartridge type)
  e) Co2 type extinguisher (Portable & Trolley mounted)
  f) Trolley mounted higher capacity dry powder fire extinguisher
  g) Dry powder extinguisher for metal fire
  h) Clean agent fire extinguisher

Pressure Testing of Fire Extinguisher

Type of Extinguisher	Test Interval Year	Test Pressure kg/sq.cm	Pressure Maintained for Minute
Water Type Gas Cartridge (IS940)	3	35	2.5
Water Type Stored Pressure (IS6234)	3	35	2.5
Water Type Gas Cartridge (IS13385)	3	35	2.5
Mechanical Foam Type Gas Cartridge (IS10204)	3	35	2.5
Mechanical Foam Type Stored Pressure (IS15397)	3	35	2.5
Mechanical Foam Type Gas Cartridge (IS13386)	3	35	2.5
Mechanical Foam Type Gas Cartridge 135 Liter (IS14951)	3	35	2.5
Dry Powder Stored pressure IS13849	3	35	2.5
Carbon Di Oxide IS2878	5	250	2.5
Clean agent IS15683	3	35	2.5
Dry Powder (gas cartridge) IS2170, IS10658, IS1183	3	35	2.5

Life of fire extinguishers

As per IS standard Annexure F (Clause 12.2.1) every extinguisher has a serviceable life. Once the life of extinguisher is completed, even though if the extinguisher is in good working condition, it has to be disposed/scrapped out.

Life of extinguisher is as appended below:

Sr. No.	Type of Extinguisher	Life time, year
1	Water Type	10
2	Foam Type	10
3	Powder Type	10
4	Carbon Dioxide	15
5	Clean agent	10

Notes:

1. Life of extinguisher s shall be considered from the date of manufacture of extinguisher

2. In case of failure in hydraulic pressure testing, extinguisher shall be discarded immediately before the life time given above.

Maintenance schedule for a Fire Extinguisher

Sr. Nos.	Extinguishers	Inspection	Maintenance	Recharging*	Hydrostatic Testing
1	Water type (Stored Pressure)	30 days (Cl.11.11 of IS 2190:2010)	Quarterly or Annually (CI.11.14 of IS 2190:2010)	2 Years (Annexure D-CL.11.4.1 & 12.3 of IS 2190:2010)	3 years (Annexure E-CL 12.2.1 & 12.2.2(g) of IS 2190:2010)
2	Mechanical Foam (Stored Pressure)	30 days (Cl.11.11 of IS 2190:2010)	Quarterly or Annually (CI.11.14 of IS 2190:2010)	2 Years (Annexure D-CL.11.4.1 & 12.3 of IS 2190:2010)	3 years (Annexure E-CL 12.2.1 & 12.2.2(g) of IS 2190:2010)
3	Dry Chemical (Stored Pressure)	30 days (Cl.11.11 of IS 2190:2010)	Quarterly or Annually (CI.11.14 of IS 2190:2010)	3 Years (Annexure D-CL.11.4.1 & 12.3 of IS 2190:2010)	3 years (Annexure E-CL 12.2.1 & 12.2.2(g) of IS 2190:2010)
4	Water type (Cartridge)	30 days (Cl.11.11 of IS 2190:2010)	Quarterly or Annually (CI.11.14 of IS 2190:2010)	5 Years (Annexure D-CL.11.4.1 & 12.3 of IS 2190:2010)	3 years (Annexure E-CL 12.2.1 & 12.2.2(g) of IS 2190:2010)
5	Mechanical Foam (Cartridge)	30 days (Cl.11.11 of IS 2190:2010)	Quarterly or Annually (CI.11.14 of IS 2190:2010)	5 Years (Annexure D-CL.11.4.1 & 12.3 of IS 2190:2010)	3 years (Annexure E-CL 12.2.1 & 12.2.2(g) of IS 2190:2010)
6	Water type 50 ltr.	30 days (Cl.11.11 of IS 2190:2010)	Quarterly or Annually (CI.11.14 of IS 2190:2010)	5 Years (Annexure D-CL.11.4.1 & 12.3 of IS 2190:2010)	3 years (Annexure E-CL 12.2.1 & 12.2.2(g) of IS 2190:2010)
7	Mechanical Foam 50 ltr.	30 days (Cl.11.11 of IS 2190:2010)	Quarterly or Annually (CI.11.14 of IS 2190:2010)	5 Years (Annexure D-CL.11.4.1 & 12.3 of IS 2190:2010)	3 years (Annexure E-CL 12.2.1 & 12.2.2(g) of IS 2190:2010)
8	CO2 Portable and Trolley Mounted	30 days (Cl.11.11 of IS 2190:2010)	Quarterly or Annually (CI.11.14 of IS 2190:2010)	5 Years (Annexure D-CL.11.4.1 & 12.3 of IS 2190:2010)	5 years (Annexure E-CL 12.2.1 & 12.2.2(g) of IS 2190:2010)
9	Wet Chemical for Kitchen fires.	30 days (Cl.11.11 of IS 2190:2010)	Quarterly or Annually (CI.11.14 of IS 2190:2010)	2 Years (Annexure D-CL.11.4.1 & 12.3 of IS 2190:2010)	3 years (Annexure E-CL 12.2.1 & 12.2.2(g) of IS 2190:2010)
10	Clean Agent	30 days (Cl.11.11 of IS 2190:2010)	Quarterly or Annually (CI.11.14 of IS 2190:2010)	5 Years (Annexure D-CL.11.4.1 & 12.3 of IS 2190:2010)	3 years (Annexure E-CL 12.2.1 & 12.2.2(g) of IS 2190:2010)

*Recharge is also required to take place after every use and if the need is identified during maintenance or inspection.

National Building Code 2016 (Fire & safety)

BIS -SP 7:**2016** titled '**National Building Code of India** (NBC **2016**)' is a comprehensive **building code** providing guidelines for regulating the **building** construction activities across the country. This code has laid down requirements for a building in view of fire and safety. The extracts from NBC 2016 Part 4 is as appended below:

Clause 3.4.6.1: For high rise buildings, following additional provisions of means of access to the building shall be ensured.

a) The width of main street on which the building abuts shall not be less than 12 meter and one end of the street shall join another street not less than 12meter in width.

b) The road shall not terminate in a dead end; except in case of residential building, up to a height of 30meter.

c) The compulsory open spaces around the building shall not be used for parking,

d) Adequate passageway & clearances required for firefighting vehicles to enter the premises at the main entrance, the width of such entrance shall be not less than 4.5meter. If am arch or covered gate is constructed, it shall have a clear headroom of not less than 5meter.

Clause 3.4.7 Mixed Occupancy

If a building is used for more than one type of occupancy, then in this case the high hazard area (such as kitchen of an restaurant) wall should of 4hour fire retardant ratings.

3.4.8.3 Openings in walls or floors which are required for passage of all building services such as cables, electrical wirings, telephone cables, piping etc. shall be protected by enclosure I the form of ducts/shafts having a fire resistance not less than 2 hour. The inspection door for electrical for these enclosure should have a 2-hour fire rating. The medium & low voltage wiring running in shafts/ducts shall either be armored type or run through metal conduits. The space between conduit pipes & the walls/slabs shall be filled by a fire rated material having a resistance rating of not less than 1 hour.

Note: In case of building where it is necessary to lower or lift heavy machinery or goods from one floor to the other, it may be necessary to provide larger openings in the floor. Such openings shall be provided with removable covers, which shall have the same strength and fir resistance as the floor.

Clause 3.4.8.4 Vertical Opening

Every vertical opening between the floors of the building shall be suitably enclosed or protected, as necessary to provide following safety:

a) Safety for occupants while using the means of egress by preventing spread of fire, smoke or fumes through vertical openings. It should be ensured that the passage/escape path has a clear height of 2100 mm.

b) Limitation of damage to the building and its contents.

Clause 3.4.11.2

Air conditioning supply treated air for more than one floor should be provided with fire dampers. The fire dampers should be designed to automatically close in case of fire incident. And should in comply with accepted standard of ((4)10). Such a system shall also be provided with automatic controls to stop the fan in case of fire, unless arranged to remove smoke in case of fire, in such case it should remain operational.

3.4.11.3 Air conditioning servicing large place of assembly (over 1000) such as malls, hotels with 100 rooms in single block, should be provided with effective system to prevent smoke circulation in case of fire. It should also have smoke sensitive devices for actuation in accordance with accepted standards. (4(11).

3.4.11.4 For fire safety, separate AHU to be provided for different floor, this is to ensure avoidance spread of fire & smoke through air conditioning duct. This should be in accordance with good practice (4(12).

3.4.12 Smoke venting

3.4.12.1 Smoke venting facilities for safe use of exits in windowless buildings, underground structures, large area factories, hotels and assembly buildings (including cinema halls) shall be automatic in action with manual controls in addition.

3.4.12.2 Natural draft smoke venting shall utilize roof vents or vents in walls at or near the ceiling level; such vents shall be normally open or if closed, shall be designed for automatic opening in case of fire, by release of smoke sensitive devices.

3.4.12.3 Where smoke venting facilities are installed for exit safety, these shall be adequate to prevent smoke accumulation during evacuation, using exit facilities with a margin of safety to allow for unforeseen contingencies. It is recommended that smoke exhaust equipment should have a minimum capacity of 12 air changes per hour. Where mechanical venting is employed, it shall be fire safe.

3.4.12.4 The discharge apertures (opening) of all natural draft smoke vents shall be so arranged as to be readily accessible is for opening by fire fighters.

3.4.12.5 Power operated smoke exhausting systems shall be substituted for natural draft vents only by specific permission of authority.

3.4.17 Skylights

3.4.17.1 Wired glass for skylights or monitor lights shall comply with the following requirements:

a) Wired glass for skylights or monitor lights-The wired glass for skylights or monitor lights shall be of minimum half hour fire resistance rating.

b) Frames & glazing- The frame should be continuous and divided by bars at not more than 700 mm centres. The frame and bar should be of iron or other hard metal and supported on a curb either of metal or of wood covered with sheet metal. The toughened glass should be secured by hard metal fastenings to the frame and bars independently of any lead. Cement or putty used for weather proofing purpose.

3.4.18 Louvers

Louvers wherever provided shall be of minimum half hour fire resistance rating.

3.4.19 Glass of façade of high rise building should be made of 1 hours fire resistance rating.

4.4 Capacities of Exits

4.4.1 The unit of exit width, used to measure the capacity of any exit, shall be 500 mm. A clear width of 250 mm shall be counted as an additional half unit. Clear widths leas than 250 mm shall not be counted for exit width.

Note: The total occupants from a particular floor must evacuate within 2 ½ minutes (90 seconds) for Type 1 construction, 1 ½ minutes for Type 2 construction and 1 minute for Type 3 construction. Size of exit dooor/exit way shallbe calculated accordingly keeping in view the travel distance as per Table 22.

4.5 Arrangement of Exits

4.5.1 Exits shall be so located that the travel distance on the floor shall not exceed the distance given in Table 22.

4.5.2 The travel distance to an exit from dead end of a corridor shall not exceed half the distance specified in Table 22, except in assembly and institutional occupancies in which case it shall not exceed 6m.

4.5.3 Whenever more than one exit is required on floor, exits shall be placed as remote from each other as possible and shall be arranged to provide direct access in separate directions from any point in the area served.

Table 22 Travel Distance for Occupancy and Type of Construction

(Clauses 4.4.1, 4.5.1 and 4.5.2)

Sr. No.	Group of Occupancy	Maximum Travel Distance Construction	
		Type 1 & 2	Type 3 & 4
		Meter	Meter
1	Residential (A)	30	22.5
2	Educational (B)	30	22.5
3	Institutional (C)	30	22.5
4	Assembly (D)	30	30
5	Business (E)	30	30
6	Mercantile (F)	30	30
7	Industrial (G)	45	-
8	Storage (H)	30	-
9	Hazardous (J)	22.5	-

Note:

1. For fully sprinkled building, the travel distance may be increased by 50% of the value specified.

2. Ramps shall be protected with automatic sprinkler system and shall be counted as one of means of escape.

4.6 Number of Exits

4.6.2 Buildings 15 meter in height or above, and all buildings used as educational, assembly, institutional, industrial, storage, and hazardous occupancies and mixed occupancies with any or the aforesaid occupancies, having area more than 500 sq.m. on each floor shall have a minimum two staircases. They shall be of enclosed type, at least one of them shall be on external walls of buildings and shall open directly to the exterior, interior open space or to an open place of safety. Further, the provision or otherwise of alternative staircases shall be subject to the requirement of travel distance being complied with.

4.7 Door Ways

4.7.1 Every fire exit door should open in enclosed stairways or a horizontal exit of a corridor or passage way providing continuous and protective means of egress.

4.7.2 No doorway shall be less than 1000 mm in width except assembly buildings where door width shall be not less than 2000 mm.. Doorways shall be not less than 2000 mm in height.

4.7.3 Exit doorways shall open outwards, that is away from the room, but shall not obstruct the travel along any exit. No door, when opened, shall reduce the required width of stairway or landing to less than 900mm; overhead orsliding doors shall not be installed.

Note: In the case of buildings where is a central corridor, the doors of rooms shall open inwards to permit smooth flow of traffic in the corridor.

4.7.4 Exit door shall not open immediately upon a flight of stairs; a landing equal to at least the width of the door shall be provided in the stairway at each doorway; the level of landing shall be the same as that of the floor, which it serves.

4.7.5 Exit doorways shall be openable from the side, which they serve without the use of key.

4.7.6 Mirrors shall not be placed in exit ways or exit door to avoid confusion regarding the direction of exit.

4.8 Corridors and Passageways

4.8.1 Exit corridors and passageways shall be of width not less than the aggregate required width of exit doorways leading from them in the direction of travel to the exterior.

4.8.2 Where stairways is discharge through corridors & passageways, the height of corridors and passageways should not be less than 2.4 M.

4.8.3 All means of exit including staircases lift lobbies and corridors shall be adequately ventilated.

4.9 Internal Staircases

4.9.1 Internal staircase shall be constructed of noncombustible materials.

4.9.2 The internal stairs shall be constructed as self-contained unit with an externa; wall of the building constituting at least one of its sides and shall be completely enclosed.

4.9.3 Staircase shouldn't be arranged around the lift shaft.

4.9.4 Hollow combustible construction is not permitted.

4.9.5 No gas piping or electrical panels shall be allowed in the stairways. Ducting may be permitted if it is of 1 hour fire resistance rating.

4.9.6 Notwithstanding the detailed provision for exits in accordance with clause 4.3, 4.4 & 4.5 the following minimum width shall be provided for staircases:

A	Residential buildings (dwellings)	1.0 m
B	Residential Hotel Building	1.5 m
C	Assembly building like auditorium, theatres & cinemas	2.0 m
D	Educational buildings up to 30 m in height	1.5 m
E	Institutional building like Hospitals	2.0 m
F	All other buildings	1.5 m

4.9.7 The minimum width of tread (A **stair tread** is the horizontal portion of a set of stairs on which a person walks.) without nosing shall be 250 mm for internal staircase of residential buildings. This shall be 300 mm for assembly, hotels, educational, institutional, business and other buildings. The treads shall be constructed & maintained in a manner to prevent slipping.

4.9.8 The maximum height of riser shall be 190 mm for residential buildings and 150 mm for other buildings and the number shall be limited to 15 per flight.

4.9.9 Handrails shall be provided at a height of 1000 mm to be measured from base of the middle of the treads to the top of the handrails. Balusters/railing shall be provided such that the width of staircase does not reduce. (see fig 1)

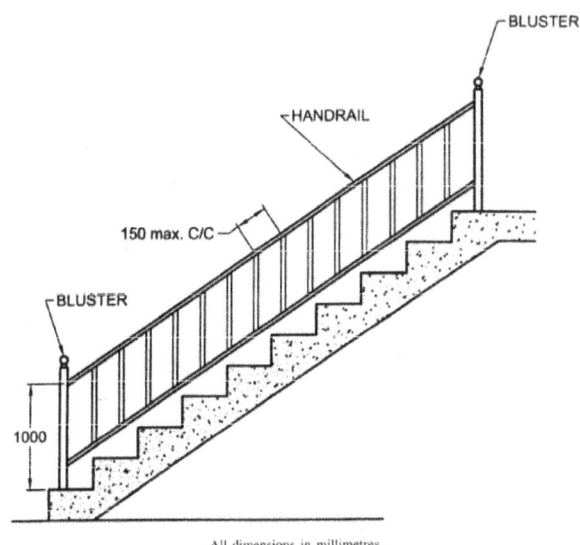

All dimensions in millimetres.

FIG. 1 TYPICAL DETAIL FOR HANDRAIL/BLUSTERS OF A STAIRCASE

4.9.10 The number of people in between floor landings in staircase shall not be less than the population on each floor for the purpose of design of staircase. The design of staircase shall also take into account the following:

a) The minimum headroom in a passage under the landing of a staircase & under the staircase shall be 2.2 m.

b) For building with 15 m height or more access to main staircase shall be through a fire/smoke check door of a minimum 2 hour fire resistance rating. Fire resistance rating may be reduced to 1 hour for residential buildings (except hotels & starred hotels).

c) No living space, store or other fire risk shall open directly into the staircase.

d) External exit door of staircase enclosure at ground level shall open directly to the open spaces or through a large lobby, if necessary.

e) The main & external staircases shall be continuous from ground floor to the terrace level.

f) No electrical shafts/AC ducts or gas pipes etc shall pass through or open in staircases. Lifts shall not be open in staircase.

g) No combustible material shall be used for decoration/wall paneling in the staircase.

h) Beams/columns and other building features shall not reduce the head room/width of the staircase.

COMMERCIAL LEASING

Commercial lease is required when the property in question is purported to be used for official or business purpose. In technical terms, commercial property is property that derives its income from non-residential sources such as offices, retail space and industrial tenants, and are bought, sold or leased strictly for their ability to produce income.

Unlike a residential lease, a commercial lease tends to cost many times more than a residential lease and is made for a longer duration.

A commercial lease is a contract between a landlord and a business for the rental of property. Commercial lease agreements are more complicated than residential leases because the terms are negotiable and vary greatly from lease to lease.

Types of Commercial Property in India:

1. SEZ

2. IT/ITES

3. Commercial

4. Industrial

5. Retail

Types of Commercial Lease

In India, commercial leases are usually term based and have a specific time limit with locking period. Apart from straight forward leases, there is another category you should know about is Master-lease/sub lease – where the lessee has the right to further lease the property to others.

- Master Lease: This is a lease agreed and signed between landlord and tenant. The original lease holder/tenant will become the Master tenant when a sublease is created.

- Sublease: The original tenant assigns some or all of his interests in the premises to a third party. The original tenant becomes the master tenant and the new tenant becomes the subtenant. The original is not replaced and is responsible for rent and damages for the term of the lease. The only way to release him from any liability is through Novation, with the landlord's consent, in the original lease.

However, in other countries there are various types of leases, and some of these (with or without the terminology) are slowly being adopted in India in case of specialized transactions.

- Net Lease: The tenant is responsible for rent plus property taxes for the premises.

- NN or Double Net Lease: The tenant is responsible for rent plus property taxes plus insurance.

- NNN or Triple Net Lease: The tenant is responsible for rent plus they pay for their share of property taxes, insurance and operating cost.

- Gross Lease: The tenant pays the landlord one set rent amount and the landlord has to make payment for insurance, real estate taxes and maintenance expenses. These are most common in multi-tenant buildings.

- Modified Gross Lease: In addition to the rent, the tenant is responsible for janitorial services provided in their space.

- Industrial Gross Lease: In addition to the rent, the tenant is responsible for paying for their share of utilities and janitorial & security services.

- Full Service Lease: All services, including utilities and janitorial & security services, are included in the rent.

- Index Lease: The amount of rent depends on a price index such as the Consumer Price Index (CPI). This is considered neutral percentage rate, because neither the landlord nor the tenant is picking the increase rate. This kind of lease is not followed in India.

- Percentage Lease: This is a percentage of gross of sales. It will vary from year to year based on the success of your business. This kind of lease is applicable in shopping malls only.

- Graduated Lease: The amount of rent for future years can vary depending on certain factors, like gross income or an annual percentage increase. This kind of lease is not followed in India

- Step up Lease: Rent is increased by a pre-set rate or set amount, to be paid on a set schedule.

- Straight Lease or Flat Lease: The amount of rent is fixed for the Lease Term.

- In case of retail, commercial, IT/ITES and SEZ property lease in India; following heads are covered under the lease.

  a) Lease fee

  b) Property Tax

  c) Parking charges

  d) Common Area Maintenance (CAM)

Precautions while drafting a commercial lease document

Due care must be taken to correctly identify the parties to the agreement along with the premises that is the subject matter of lease agreement. The space the tenant leases is known as 'premises' and it becomes important for both the tenant and the landlord to clearly define the precise square footage of the premises.

Most of the promotors or developers who lease their premises mentioned offered area for lease by various names such as super built up area, chargeable area or built up area. The super build up area means addition of common area in proportion to the carpet area.

Super Built Up Area

Lift lobby, main lobby, terrace, green areas, pathways, shaft, green area, machine room, corridors, all staircases, shaft including lifts, security room, common washrooms, common AHU rooms are covered under super built up area.

Tanks (water, diesel), roads parking area, ramps, terrace, 50% of open balconies are excluded from super built up area.

Due to above reason of inclusions the premise defined in the lease often does not match the premise that the tenant expects to receive. Thus extra care must be taken to ensure that there exists no ambiguity in this respect. This is most important when you are entering into a lease contract with the builder himself – in case of a mall or office building. The space you are actually supposed to get should be mentioned in the agreement in an identifiable manner. Normally the promotor/builder charge for super built up area and they don't quote as per carpet area which a tenant will be actually using.

Hence it is always advisable to get quotes on carpet area only and before signing the lease deal the carpet are should be physically measured and cross checked.

As per IS 3861:2002 area definition is Super built up, Built up and Carpet area.

Types of Spaces

1. Bare shell

2. Warm Shell

3. Furnished

Bare shell: Bare shell kind of space, is available for lease but doesn't have any furnishing, firefighting & alarm system, sprinkler system and air conditioning system. Those who leases this kind of space need to use their own capital funding for interior fit out including air conditioning system.

Warm Shell: Warm shell kind of space is as good as bare shell kind of space, the only difference is in warm shall the developers or landlord provided chilled water tap off point from chiller or has provided basic facility such as lighting, air conditioning and toilets. The lessee only has to install AHU & associated ducting system and furniture's. The building will be having centralized chiller plant. Also in some cases the landlord provides finished toilets and lift lobbies.

Furnished: This kind of office space is ready for use. The lessee has to just get their PC's and files and they can start operating from this kind of space.

There will be minor difference in lease fee between Bare shell and Warm shell. However, the lease fee for furnished office space will always be higher than the other two kind of spaces.

Lease Term and Commencement

The term of the lease is the time period during which the tenant has exclusive possession for usage and obligation to pay rent for the premises. The dates of commencement and termination of the lease is required to be mentioned in the lease deed. The tenant's needs must be known in advance to determine the length of the lease.

The standard market practice is a minimum of 05 years lease with 03 years locking period for tenant and 05 years locking period for landlord. Lease tenure can be more than 05 years it totally depends on the tenant's business requirement. However, it is advised that locking period should never be agreed for a term more than 03 years by tenant.

Note: Locking Period is a period in which neither the landlord nor the tenant can terminate the lease agreement.

Most of the times, tenant have to make capital investments of substantial value in making the leased space operational, these spaces can be a manufacturing unit, office fixtures, interior decoration. And it makes sense only if the lease is sufficiently long term. A minimum of 05 years of lease will help, but it depends on individual company's policies and kind of capital being invested. This should be addressed in the agreement, and if the owner of the property wants to terminate the lease prematurely leading to losses, such losses should be calculated and recovered under the agreement by the tenant from property owner. This can be ensured by having a locking period on landlords end also. This locking period can be 05 years or more, depending on negotiations.

The date of lease and its commencement date can be different. If you are moving into a furnished office space then the dates can be same. But if you are moving into warm shell or bare shell then both these dates have to be different. The commencement date should be from the date of completion of improvements that the landlord agreed to or date of completion of fit outs by tenant. Hence, if the lease date and commencement date are different then the tenant must ensure that the improvements (if any) must be finished as per his requirements before paying the rent and also make clear whether tenant's non-rent obligations begin before the commencement date to prevent any sort of problem.

The fit out period: When a tenant leases a warm shell or bare shell office space or any land is leased for a workshop or factory, in this case the tenant company has to undertake interior fit out or construction works to create an office space with furniture and associated required systems or setting up of a workshop/factory. Depending on the size of area leased the fit out period is decided. The fit out period can be anything between one month to three months. This is a matter of negotiation and negotiation skill of the administrative/Facility Manager. In some cases the greater fit out period was negotiated by tenant.

Fit out period is also known as rent free period. During this period, the tenant is not required to pay any rent.

Rent/Lease Fee

The rent forms a basic element of the lease agreement and must be clearly discussed by the tenant and the landlord. The mode of payment, the rate at which it will increase, due date etc. must be clearly stated in the agreement.

As per the market standards a normal lease is signed with three years locking and total lease can be for 05 years. During first three years there is no lease fee hike after three years the lease fee hike is applicable as 15%. And from date of hike in lease fee till expiry of lease agreement there will not be any hike. If the corporate wants to extend the lease they need to sit again with landlord and negotiate. Most of the time looking at market conditions landlord agrees to extend the lease with same terms and conditions without any hike.

There is another way of negotiating a lease, where if you have hard negotiated on pricing with the landlord you can offer the landlord 5% hike every year. With this arrangement there will be approx. 16.5% rise in lease fee after three years. This will be a kind of win-win negotiation.

Security Deposit: In order to ensure reliability from lessee's end, the landlord usually asks for a significant sum as security deposit. The security deposit asked are equivalent to lease fee ranging from 03 months to 12 months. One has to ensure that a correct mention of security deposit is made in lease agreement. Security deposit is always considered as interest free and can be adjusted against the dues on tenants.

Security deposit is also a matter of negotiation. Tenant can negotiate less lease fee against the higher security deposit.

A monthly due date for rent payment should be fixed, which is usually 7th of every month, but the parties are free to determine any date. The grace period before a penalty is charged must be subject to negotiation. The penalty amount must also be discussed. The standard practice of interested on late payment of lease fee is 1.5% per month i.e 18% per annum.

The tenant must carefully review the method in application for the rent increase during the term of the lease agreement. They should ensure that the increases must be based on the Consumer Price Index or a fixed percentage. The standard lease fee increase for commercial property is considered at 15% after three years, or one can also negotiate an annual increase between 4% to 5%.

Expenses and Common Area Maintenance (CAM)

A CAM (Common Area Maintenance) charge is an additional cost, charged on top of base rent by the building management, and is mainly composed of maintenance fee for work to be performed on the day to day upkeep of common area and common services of a property. While negotiating a lease agreement one should review the expenses the landlord asks you to pay as CAM and make sure that they are reasonable, and related to building's operations and consistent with the lease agreement.

Certain costs, like the landlord's staff's salaries, capitalized expenses (including major repairs), taxes on income and costs related to landlord's violation of any governmental regulations are not the tenant's responsibility and should not be allowed to creep into the agreement.

Common area maintenance (CAM) charges are certain percentage share of the costs of the landlord's for maintaining, operating, replacing and repairing the components of equipment's installed in the building (eg. lift for the building or common area electricity usage), cleaning services, security services etc. which the tenant must shell out for. Thus all CAM charges must be pre-determined and agreed upon by both the tenant and the landlord.

Repairs and Maintenance of common area premises and common services is responsibility of landlords and tenant is only responsible for keeping the leased premises in good and safe condition. This is again a point of discussion during lease negotiation, if the complete building/premises is lease by a single tenant, the tenant can accept responsibility of Common area maintenance or can ask landlord to maintain and bill as per actuals. In case of multi tenanted building it should be clear that the building maintenance and common services has to be maintained by the landlord.

However, in some cases these obligations may be evenly divided between the landlord and tenant. If so, it is needed to specify clearly which party is responsible for what repairs and maintenance. It is a good idea to put a cap on the expenditure the tenant must incur for such repairs.

The list of expenses as appended below, associated with the Repair, Replacement, Maintenance & Operation and other services needs to be taken into consideration for assessing the CAM fees/ rate (applicability of each will depend upon the type of premises being managed):

- AMC for Common utilities which mainly includes Transformers, Diesel Generators, its Distribution panels & Circuit breakers, Air Conditioning System, Emergency Inverters/UPS, drinking water & Wastewater treatment plants, Firefighting System.

- Repair & Maintenance expenses involving common utilities such as HT & LT breakers, Diesel Generators, Inverters, CCTV System, Capacitor banks, Batteries, Firefighting and alarm system, Public address System and Portable Fire extinguishers), and various pumps & motors (Submersible for Bore water, water bodies & sump water, Hydro- pneumatic for water tanks, washroom exhaust & fresh air blowers).

- Stores & Consumable expenses involving Electromechanical, Plumbing, Electrical, HVAC spares; Paint, Oil & Lubricants, Door closers, Floor springs, Glass sealants, Gaskets etc.

- Civil works expenses involving repair & maintenance such as driveway paving blocks, speed breaker, barricades, fabrication work, wall cracks/ leakage repairs, replacement of damaged façade glass etc.

- Property Management fees & salaries for Electrical, Mechanical, Plumbing, BMS, Housekeeping, Security, Help desk & the management staff.

- Electricity/water charges in respect of the common areas.

- Other Miscellaneous expenses involving periodic Façade & Water tanks cleaning, Upkeep of Horticulture/ gardening/landscaping/potted plants, Office infrastructure, Cost of equipment and tools, Property taxes, Property Insurance costs (excluding the Licensed Premises), Waste management & disposal, Promotion and Marketing charges (for Malls)

- Compliance and Statutory payments (Various Meters calibration fee, Lift Inspection fee, Electrical Inspection fee, Audit fee, DG annual fee), GST, property tax etc.

- Management fee

- Sinking fund (Depreciation cost of equipment's)

How to Calculate CAM Charges?

There are various ways to calculate CAM charges.

The total leasable square footage need to be divided by tenant's square footage. For example, if the tenant leases 10,000 square feet and the GLA (Gross Leasable Area) is 1,000,000 square feet, the equation looks like this: 10,000/1,000,000 = .01. Multiply by 100 to change the decimal to a percent. This percentage represents the proportion of the **CAM** fees to paid by that particular tenant. This calculation will be based on actual consumption. And this will keep varying every time a tenant moved in or moved out.

The other way of fixing a CAM charge is that you need to budget on all above cost, divide it by GLA and you will find CAM charges per square feet.

Normally the developer or building management team, budget the CAM charges and make it a fixed cum variable component.

The standard CAM charges as on date vary from INR 15 to 20 per sq.ft. per month, depending on standard and grade of buildings.

Subleasing of leased property

Subleasing is required only when a company is not doing good in business and they are looking for reduction in overheads. One of the option for reduction in cost is facilities cost, by surrendering some of its space. However, if it is not allowed to be surrendered due to binding of locking period then the only left out solution is sub leasing. In today's scenario all of the top developers don't permit sub leasing and they ensure to add this clause in lease agreement.

The tenant must know their rights to sublease/ assign the premises. Often there are strict prohibitions on subleasing or assigning the premises unless the landlord's consent is obtained but whether it is to be granted or withheld is entirely the landlord's prerogative. Many safeguards can be introduced in the lease deed in relation to this to prevent the landlord to use this power against the tenant. These may include creditworthiness or financial strength of the proposed assignee or demonstrating that the proposed tenant does not violate any exclusives that the landlord has set. Further, the tenant may also add a provision in the agreement setting a time bar on the landlord to give consent and any delay will waiver the landlord's right to object. It is a matter of negotiation with the developers. If the lessor is an investor and one may succeed into having an agreement on subleasing.

Non Use or Misuse of Property

In case of commercial properties, in the event of non-use for more than 6 months of the premises can give a right to the landlord to file a suit for eviction against the tenant under certain laws in India. Reference, Section 13 of Rent Control

Act of Bombay. There is also a Public Premises Act for premises leased or rented by public or governmental bodies. In case of any misuse or lack of use, or even use which is not covered by the lease agreement (such as utilization of office space for storage or warehousing), the government body may be able to evict the tenant on basis of this Act.

Compliance with Applicable Law

The provisions in many leases required tenant's to be compliance with 'all applicable laws' during the term of the lease may seem innocuous at first. What could be unreasonable about a requirement to comply with the law? Such a clause in the agreement may make the tenant responsible for matters that are already not in compliance at the commencement of the lease term. This could be fault of a previous tenant or the landlord, and the new tenant should be made to suffer the consequences.

Consider a case where a company in chemical manufacturing has vacated a premises and had its process has affected the environment such as contamination of ground water, contamination of earth etc. etc, and another company leases the same space then the liability will be of existing tenant.

This should be specifically addressed in the agreement. In such situations the tenant may either agree to comply with all applicable laws relating only to the tenant's use or business at the premises during the term or they could obtain a representation and a warranty from the landlord that premises are in compliance with all applicable laws as of the commencement date. Taking such a representation is a good idea in any case.

Premature termination

While discussing Landlords ask for a right to prematurely terminate a tenant's lease, by way of notice and or in case of breach. However, this is a very sensitive issue for the tenant, as this can lead to huge losses. One must ask for a minimum of 03 to 06 months of notice, and if possible entirely exclude such a possibility.

Many commercial landlords are often fine with giving up this right in return of small amounts, which is a good idea for a commercial tenant. At least, the tenant can ask for a period during which the landlord cannot exercise this right. Even the scope for termination for breach should be limited as far as possible as it is a serious commercial threat. Mutual terminations should be fine at any time.

It is always advisable for locking period to be agreed for both tenant and landlord. Tenant have a locking period of 03 years and landlord can be asked for a locking period of 05 years. In case of premature termination either party becomes liable for a penalty equivalent to the lease fee of balance period. For example, if the tenant is bonded with 03 years locking period and they want to move out of property after 2.5 years, then the tenant need to pay balance 06 months' rental as penalty and same will be applicable for landlord if he/she asks tenant to vacate the premises.

Insurance Matters

Insurance is the responsibility of both the landlord and tenant. Landlords are responsible for insuring all his/her asset against all kind of threats & with right valuations. Tenant will be responsible for insurance of their own lease premises and assets. Insurance matters in the lease agreement must be diligently reviewed by both the tenant's and landlord's insurance agent, if any, to make sure that:

- They are reasonable
- They can be satisfied.

Business/Operational Needs of the Tenant

The tenant must be clear about all their operational needs and must include terms that ensure his ability to carry out his business without hindrance, successfully and to his satisfaction. Such operational needs include parking space, location of the premises, environment around the premises, availability of electricity, cooling/heating services, security etc.

Tenant should also ensure to add a clause of peaceful conduct of business in the premises and ensure safeguard against any disturbance to their business and in case of disturbance they will have a right to move out of lease agreement. Disturbance to the business can be due to surrounding environment of the premises, uneasy access of employees to the premises, landlord not paying electricity dues or property tax or defaulting on his loans and bank sending a notice to the tenant.

Damage or Destruction of Premises

In case of damage/destruction of premise, many leases contain a very pro-landlord provision leaving the tenant to adjust with the repairs being made which may not be to his liking or requirement or he may not know how long the repairs will take. This will adversely affect his business in all respects. Hence there should be a time limit set by the tenant wherein the landlord must complete all repairs. The standard practice is between one month to three month of time given to landlords for repair and restoration of building post that tenant will have right to terminate the and move out of premises.

Relocation Rights

If the first draft comes from a landlord, especially mall owners and major developers, there may be clauses that give the landlord the right to relocate the tenant to a new location within the building or locality. This affects the tenant's business as location is a significant factor in building the repute of a business. It is not a standard clause in India and most parties manage to remove these clauses successfully through negotiation.

Thus the above mentioned are certain issues that the tenant must keep in mind before entering into a commercial lease agreement and ensure that his best interests are realized to be able to expand his business. Knowing these issues in advance will ensure the most effective representation of the tenant's business.

Search for Commercial Property

Before you start search for a leasing a commercial real estate, you need to have few answers available which will define a properties parameter. This is because there are a wide range of commercial properties available for businesses of all types. These parameters will help you limit your search to commercial spaces that suit your needs. You need to have answers to below quarries:

1. Customer location: It is a standard practice to have your office location close to customer's office location.

2. Employee pool: The location should have availability of required talent (personal). Example: Most of the IT MNC's prefer office space in Bangluru or Pune, just because of availability of talent pool.

3. Property type and zoning: property type can be industrial, warehousing, IT/ITES, SEZ, Commercial etc. All commercial real estate properties are zoned for a specific use. A warehouse is a good example of a commercial property zoned for industrial use. Other commercial zoning includes leisure, office, retail, and restaurant. The type of zoning dictates the type of business that can operate out of the commercial building. For example, if

you're looking for office space, you won't be able to lease property zoned for retail or a restaurant. Conversely, you can't rent a space zoned for offices and convert it into a restaurant. Make sure that you understand your local zoning laws as well as the type of zoning your business needs. Further you can even can't lease office space in a ITeS space if your business is commercial.

4. Desired size: The commercial lease options available are largely dependent on the size and layout of the space you need. To calculate the size, you'll need to determine the number of customers or the size of your workforce in order to derive the necessary square footage. For example, restaurants and retail locations typically require 15 square feet per customer on average. Offices, on the other hand, typically require between 100 – 150 square feet of usable workspace per employee. In case of BPO it can be 50 sq.ft. per person, similarly one need to know about the requirement from the business or a standard being followed by your companies' policy.

5. Maximum budget: This can be discussed and approved by the Management before finalizing a property. Another thing you'll want to determine is your maximum monthly budget. This will help you limit your searches to only spaces that you can afford. The maximum budget is largely dependent on your business's size and performance. To help, it's important to determine the average price per square foot for your area.

   Typically, landlords share price per square foot on super built up area (Gross Leasable Area). However, this costing doesn't give correct picture of cost. GLA lease fee vary from developer to developer and the efficiency offered. Efficiency on GLA varies from 50% to 100%. Efficiency on GLA is mainly available from 60% to 70%, you will hardly find GLA above 70%.

   GLA (Gross Leasable Area or Super Built Up Area) efficiency mean, how much carpet area you will get. Example: for a GLA of 10,000 sq.ft. if an efficiency is considered as 60%, then in this case you get a carpet area of 6,000 sq.ft. While budgeting one should always ask for pricing on usable carpet area and not on GLA. Once you find the average price per square foot, you can take it and multiply by the square footage you need for your business. This should give you your expected annual budget for your commercial lease. You need to also add expected utilities and common area maintenance fees (CAM) and include it in your max budget calculations. You'll also want to include expenses for rent increase. For rent increases, expect that your lease payment might increase as much as 15% after every 03 years or 5% to 6% annually depending on your negotiation skills.

   From there, you need to ensure that your maximum budget doesn't exceed 8% of your expected annual gross income. Anything higher could put your business in financial distress.

6. Accessibility: Accessibility is another important factor, such as proximity to airport for easy travel of your customers, accessibility to public transport system and housing complex availability in nearby vicinity for your employee's effortless movement. Accessibility is also a major parameter for retail businesses and restaurants. For example, these businesses will want to have adequate parking for their customers. Further, they'll want to choose a location with a high amount of foot traffic and vehicle traffic.

7. HVAC & Electricity Billing: You should check with developer how the HVAC (in case of centralized air conditioning) will be billed. If the landlord uses BTU meter for billing, then you should go ahead with deal and if landlord uses other wage way to bill than you should reconsider your plan to get into lease with such landlords. Similar is the case with electricity consumption billing. Landlord should provide a dedicated meter and should agree to get it tested and calibrated every year.

8. Parking: Before finalizing a lease you should ensure that the required number of parking is available with landlord. Normal standard ratio of parking is 1:1000 or maximum 1:1500. It means 01 car park per 1000 sq.ft. of super built up area or Gross Leasable Area.

Commercial Real Estate Broker

It is always advisable to facilitate real estate leases through a good and authorized brokers. Most of the MNC's have they empaneled brokers with a global mandate. In case your company don't have such arrangements then before taking services from any broker you need to have a proper due diligence in place. There are typically two types of commercial real estate brokers involved:

1. Leasing agent – Brokers who represent landlords

2. Tenant broker – Brokers who represent tenants

Listing agents are hired by a landlord to list their commercial property. Listing agents gets a commission equivalent to one or two month's lease fee depending on the agreement, which is paid by the landlord.

Tenant brokers, on the other hand, represent tenant interests. However, tenant brokers also typically earn a percentage of the overall commission paid by the landlord, knows as the tenant broker's fee or they ask for a brokerage from tenant again equivalent to one or two month's lease fee depending on an individual's negotiation skills.

Logically the tenant broker should work towards safeguarding tenant's interests, however this doesn't happen in most of the cases. My personal experience is bad, most of the brokers are only interested in closing the lease deed and in getting their fee from both end. It is always advisable; as a tenant you need to do your own homework on market and you should be good in negotiations. Never ever rely on inputs shared by a broker.

As the brokers get their fee from both side, you can negotiate with brokers on fee and they will be more than comfortable to agree to work on zero brokerage.

Tenant Broker's Responsibility

It's not mandatory that a tenant uses a broker. However, tenant brokers can typically help a tenant in the following ways:

- List of available real estate

- Market scenario

- Accurate market pricing and comp data

- Knowledge of local market conditions

- Access to financing options if required

Further, this value is usually free to the tenant since the landlord typically covers the tenant broker's fee. It's therefore usually a good idea to engage a tenant broker and have them help you find suitable locations to lease.

Shortlisting of Commercial Broker

I needn't to mention here how to find a commercial broker. It is always advisable to take services of established and experience brokers who have their own team. In India, JLL, CBRE, C&W, Knight Frank are few names to mention, however there is a long list of brokers available in the country. Before finalizing a broker, its better you ask the few questions such as:

➢ What is the broker's experience with your specific commercial needs?

➢ What is the size of the broker's real estate practice?

➢ How is the broker being compensated?

➢ What is the broker's fiduciary duty?

➢ Is the broker knowledgeable with the local market?

➢ Is broker having ethical values?

You should ensure that the shortlisted broker has the right mix of experience and attention for your need.

If you have a broker that's too successful, you may be his/her low priority. If you choose an inexperienced broker, you may end up paying more or having a lease least favorable to you. It is always advisable not to choose a broker but instead choose a team. Pick a junior/senior combo so that when you're hunting for space you work more with the junior, and when it comes to negotiating the deal you have the experienced veteran leading the negotiation."

Once you've identified a broker you trust, you'll have to sign a written contract. This contract usually stipulates that the working relationship is either:

• Exclusive arrangement

• Nonexclusive arrangement

Let's take a look at the differences between the two.

Exclusive Arrangement with Commercial Broker

An exclusive arrangement is one where the tenant wants to work exclusively with one broker for a specified term which can be 3 to 12 months. During this time, the tenant can't work with another broker. A commission between the tenant and broker is negotiated, equal to a month or two-month lease fee as broker fee. However, this is not the right way of finding a suitable commercial space.

This, of course, is rare and landlords will almost always pay a tenant broker fee, thus waiving the commission between the tenant and broker. This is a good option because the tenant broker will then have a fiduciary duty to the tenant.

Nonexclusive Arrangement with Commercial Broker

A nonexclusive arrangement comes in two forms: 1) right to represent and, 2) not for compensation.

Right to Represent Nonexclusive Arrangement

A right to represent nonexclusive arrangement is similar to an exclusive arrangement except that a tenant is allowed to speak with other brokers. However, the tenant still pays a commission, due even if a lease is signed with through a broker.

Not for Compensation Nonexclusive Arrangement

A not for compensation nonexclusive arrangement, on the other hand, gives a tenant maximum flexibility. It's nonbinding and there are no commissions negotiated. Instead, it gives the broker the right to speak on your behalf and schedule listings for you to see. However, while it provides flexibility, this arrangement gives the tenant broker less of a fiduciary duty.

Saving on Brokerage Fee

If you have no budget to be paid as brokerage fee, there are two ways you can make an approach.

a) You can start search for a commercial real estate property yourself, you can find commercial real estate listings on internet or at time even the landlord leasing team or broker may directly contact you if you spread a message in market. However, if you choose to look for yourself, you'll have to conduct the following without the help of an experienced broker:

- Finding new listings

- Setting up walkthroughs

- Negotiating the lease

The only benefit of not using a broker is that there's no chance of paying a commission. Otherwise, it's probably best to use a tenant broker.

b) Many brokers if given an opportunity are readily available to help you with property search without charging any fee. They will help you with listings, can set up site visit and arranging meeting for negotiation.

Types of Commercial Leases

There are three types of commercial leases. The major difference between them is the way costs and fees are assessed.

The three types of commercial leases are:

- Full service lease

- Net lease

- Modified gross lease

Full Service Lease

In a full service commercial ; the lease fee expenses associated with the property, includes property tax, insurance, repairs and maintenance, utilities, security and janitorial services. In this kind of lease all the services are being managed and paid by the landlord. In this kind of businesses one can forecast their monthly and annual lease payments without any variance. In this scenario, the landlord takes on the responsibility of maintaining the property.

Net Lease

In a net lease agreement, the landlord charges lease fee which shall include property taxes, property insurance, and common area maintenance items (CAMS). Net leases can be either a single, double or triple net lease.

In a single net lease, a tenant pays rent plus a pro-rata share of the building's property taxes.

In a double net lease, the tenant pays a portion of the property insurance in addition to rent and property taxes.

In a triple net lease, the tenant pays the pro-rata share of property taxes, property insurance, and CAMS.

This means that while the base rent is lower for the tenant, the tenant is also responsible for the monthly costs associated with maintaining the property. These expenses are typically added onto the base rent monthly. In a triple net lease, you'll still need to have property insurance.

Modified Gross Lease

A modified gross lease compromises of the full service lease and the net lease. With a modified gross lease, a tenant might pay for portion of their property taxes, property insurance, and CAMS, but they pay it as a lump sum payment along with their rent.

The rent on a modified gross lease is therefore fixed and there are no hidden costs or unexpected charges. If any of the taxes, insurance, or CAMS increases, the rent remains the same. This is not the case with a net lease. For common area utilities and janitorial services are covered by the landlord with a modified gross lease.

How to Identify Right Commercial Property

When considering different commercial property listings, make sure you assess the following in addition to your property parameters:

- Location – Make sure that the property is either around your ideal customer or ideal workforce, close to public transport and cheaper housing availability. You should also ensure to check for adequate parking for customers or employees.

- Amenities and Services – You'll want to understand the full range of amenities offered by a commercial space. These amenities and services may include such things as the building is certified Green Building, it has pray area, crèche facility, common dining options, outdoor open and green space, sewage treatment and waste treatment plant, utilities, power backup, on-site security-Physical & Electronic, and more.

- History of Landlord – This is important to understand since commercial leases are typically multi-year agreements. The landlord you choose will most likely dictate the lease agreement, changes to the agreement, rental increases, and more. Your comfort of future stay will totally depend on the behavior and background of the landlord.

Property Visit & Inspection

To start with search of office space, you need to ask your broker to share multiple suitable options as per your preferences. The broker will end up sharing and exhaustive list of properties, which may be suitable or may not be suitable for your requirement. Have a one on one discussion with your broker, understand the list of properties and choose and prepare a list of properties which suits your requirement and can be visited for inspection. This discussion and analysis will help you get a better understanding of average price and gives you a leg up during the negotiation process. During your search you'll also want to compare rents to each other to ensure you're staying on budget.

A rule of thumb is that you should consider between 4 to 10 commercial properties prior to signing a lease.

Example for comparison:

	Proposed Office space in		
	Property A	Propert B	Property C
Area in Sq.ft.	16000	16000	16000
Proposed Rate from Landlord INR	250	98	70
Proposed Rate from Landlord $	3.57	1.40	1.00
Expected rate on Negotiation INR	220	100	85
Expected rate on Negotiation $	3.14	1.43	1.21
Common Area Maintenance INR	25	16	20
Common Area Maintenance $	0.36	0.23	0.29
Property Tax INR	2	2.5	2
Property Tax $	0.03	0.04	0.03
Car Parking Charges for 08 cars in INR	48,000	45,000	1,00,000
Car Parking Charges for 08 cars in USD	685.71	642.86	1428.57
Monthly Rent in USD	50,285.71	22,857.14	19,428.57
Monthly Common Area Maintenance in USD	5,714.29	3,657.14	4,571.43
Monthly Car parking Charges in USD	685.71	642.86	1,428.57
Monthly Property Tax in USD	457.14	571.43	457.14
Expected Total Monthly Outgoing	57,142.86	27,728.57	25,885.71
Capital Requirement in USD	1,75,000	2,57,143	2,57,143

Negotiations

Once you have taken a decision on commercial property options than you should start to negotiate the leases. Before entering into a commercial lease negotiation process, you should ask the landlord or its leasing team to share term sheet with you.

Term sheet basically contains term and condition of the lease. A sample is as appended below:

Sr. No	Particulars	Commercials	Note
A	Financials		
1	License fee per sq.ft. on super built up per month		
2	Interest Free Security Deposit		

3	Refund of Security Deposit		
4	CAM Charges per sq. ft. of Super built up area		
5	Interest Free Security Deposit against CAM		
6	Electricity Deposit		
7	Escalation after 03 years		
8	Property Tax		
9	Car Parking charges		

Sr. No	Particulars	Commercials	Note
B	General		
1	Duration of lease (Year)		
2	Termination/Renewal Notice period (months)		
3	Lock in period (months)		
4	Type of agreement		
5	License commencement Date		
6	License Fee Commencement Date		
7	Power Backup		

Sr. No	Particulars	Commercials	Note
C	Others		
1	Total Leasable Area (Sq.f.t) on which lease fee is payable		
2	Office Hours		
3	Signage Location		
4	Billing Cycle		
5	Payment Term		
6	Validity of term sheet		

Once you agree on term sheet, the same is signed and accepted by both parties, you can go ahead with issue of letter of intent (LOI) that represents your offer can be signed and accepted by both parties. The LOI is a chance for you to sell the landlord on why you would make a great tenant. This is especially beneficial in a commercial real estate market with high demand. Once LOI is signed and accepted by both parties, then they are bonded by the terms and condition of LOI, until a lease agreement is signed and accepted.

Letter of Intent will have a validity of one month or as decided by both the parties. Post expiry of validity the LOI becomes null and void. And either party can back off from their agreed term and condition.

Some Important Points to Ensure

Regardless of the type of lease, commercial leases will often have similar lease terms. While the payment structure might differ, all leases include such things as the required deposit, the length of the lease, and more. Let's understand terms of lease:

- Use clause – This clause determines the type of business that can use the space. For example, some spaces are zoned for retail while others are zoned for office spaces. This use clause is particularly important if you expect to sublease your space in the future, since it limits the businesses available to sublease.

- Length of lease – Commercial leases typically range from 3 – 9 years, with escalation and renewal clause every 03 years. A short term lease is always beneficial for the business because it gives a business flexibility and reduces any future financial burdens.

- Assignability – A lease has to be "assignable" if a business wants to eventually sublease the property. However most of the developers don't accept sub lease clause in the agreement. Still you should push for an assignable lease if not agreeable for sublease but should be able to accommodate for sale of a business. For example, a company if bought out by another company.

- Capital expenditures – These expenditures determine who's responsible for the repairs, maintenance, and other costs associated with the commercial property. A net lease, for example, charges a tenant for all the capital expenditures. A full service lease, on the other hand, requires the landlord to cover all capital expenditures.

- Rent and escalation – All leases stipulate not only the monthly and annual rent but also any future rent escalations. An escalation is a term that allows a landlord to legally increase the rent during the lease. It's common to see rent escalations equal to 15% every 03 year or 5% every year. One should ensure that there are no abnormal escalations in the lease

- Deposit – Most leases require a deposit. The deposit is fully refundable and protects the landlord from a delinquent tenant or a tenant that causes excess damage to the property. The typical deposit is between 3 – 6 months' of rent.

- Built to Suit Lease build – In this case landlord will complete the interior fit outs with furniture and will also maintain the facility. In lieu of same the tenant will have to pay an enhanced lease fee.

- Termination clause – Clause within the lease that allows the landlord and/or the tenant to terminate the agreement under certain conditions. Termination clauses are great if it allows you to terminate the agreement, but increases your risk if the landlord can also terminate the agreement.

- Rent abatement – This term stipulates that if the commercial property is damaged, the tenant won't have to pay rent (or pay a reduced rent) until the damage is fixed. This is a great way to reduce a business's risk.

Point to ensured when discussing lease term and area

The developer when discussing the lease term will never share correct information's with you; such as, the carpet area mentioned by the developer doesn't only include the floor area within the wall it also includes area in passage, lift lobby & wash rooms.

The gross up factor is the difference between the square footage of your actual useable area and the listed rentable area. This difference is a tenant's proportionate share of common areas.

For example, an office space of 10,000 square feet of useable area usually is measured at 12,000 square feet. However, you will be actually using 10,000 square feet. This is over and above the chargeable built up area (Gross Leasable Area)

When looking at your lease, you'll want to negotiate terms so that you're minimizing your gross up factor. However, all the developers use this method to calculate carpet area and GLA. So you have no choice left here except to bargain on lease fee and other terms.

With my experience, I have always found developer including toilet, lift lobby, passage lobby in carpet area. And they add staircase, parking area, building lobby, drive ways on super built up area or gross leasable area.

Lease vs Buy Analysis

An organization should have a process in place where they need to do a lease vs buy analysis. If the analysis output is to buy the property, then they need to buy if the corporate has the required funding in place and if the output is lease then only they can lease the property. There are number of factors considered for lease vs buy analysis such as fair market value of the property, lease term, appreciation of property, rent escalations.

After looking at a number of spaces that are available to lease, you may be confused whether it's better to buy or lease commercial real estate. Some of the top developers don't sell their property. In my view it is always better to purchase commercial real estate rather than leasing it, if you have a stable business, have required fund readily available and you are assure that you need the property for longer duration. For example, if you buy a commercial space, you'll take advantage of equity, depreciation, cash flow, and asset appreciation.

Advantage of buying commercial property over leasing:

- Build equity: You can use this equity as collateral for additional expansions.

- Appreciation: Over a period of time commercial properties market value increases, letting you eventually sell for a profit. However, assets can also lose value, making it an investment with risk that you need to consider.

- Depreciate the building: You can claim the annual depreciation on your tax returns.

- Increase cash flow: In case in future if your business is reduced than you can lease out the unused area and receive rental income.

When it comes to the benefits of leasing, tenants get to avoid any capital investments. Instead, they pay a refundable deposit, which is often far below than the required down payment for commercial loan. The difference between lease security deposit & loan down payment can be approx. 20% to 30%.

Further, lease payments can be deducted, reducing a business's tax burden. This is in contrast to owning a property, which only allows you to depreciate the asset over its useful life. Of course, if you finance a commercial real estate property, you can also deduct interest payments and origination fees.

Co-Working Space

This kind of office space, are furnished and loaded with all the services required for operations of an office. A number of companies shares these premises. This kind of space is suitable for short-term requirement or by a startup or small company who doesn't want to spend on lease of a bigger office space than required, office management and other services, but like to operate from a decent office space located in a decent area.

There are challenges for getting a small and decent office space may be for one seater or five seaters. Also leasing an office space will need capital investment and long-term commitment. To avoid any kind investment and long-term commitment with flexibility to reduce or increase in number of seats; co-working space is the best available option for such requirement without any commitment.

The co-working space will have workstations, washrooms, tea/coffee vending machine, drinking water, printers, internet connectivity; air conditioned space, meeting room, dining space, dinning space, lounge etc.

The service charges include charges for all above mentioned services however there will be limited availability of meeting rooms and printing services.

Note: One should check all documentation from the property owner before signing any lease. Please do not believe developers or property owners on verbal commitments.

COLOR CODING OF SERVICE PIPE LINES IN FACILITY

Pipes are used in facilities to transport liquids and gasses from one place to another, both short and long distances. Most facilities have dozens of pipes moving different things around the facility, so it is important to be able to keep track of what is in each pipe and where it is going. This is where pipe color-coding can become essential.

If pipe color codes are not used, it can be difficult to know what is within a given pipe, which can present many dangers. If someone opens a pipe that is not marked with the proper colors, he or she may expect clean water but get a toxic chemical.

Pipe color-coding is not a complicated process, especially if industry standards are used. There are many standards out there from a variety of sources, but the most popular is the ANSI/ASME A13.1 standard. This standard explains colors, text, size, and placement of pipe marking labels.

When using pipe markings, a facility must choose to either buy generic pipe markings, special order custom pipe markings, or obtain blank pipe markings that can be printed on using an industrial label maker. Each of these options can be effective. For facilities that have lots of pipes, the label maker is going to be the ideal option. For those who need just a few labels, special ordering is likely the best choice.

- Yellow with Black Lettering - This option is used for any pipe that contains flammable liquids and gasses. This could include gasoline, oils, and many others.

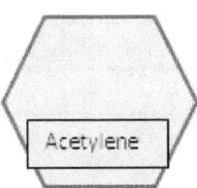

- Green with White Lettering - Using a green label with white lettering means that the pipe contains potable water. It could be used for cooling, feeding boilers, or even going to drinking fountains or sinks.

- Blue with White Lettering - The blue labels with white lettering are used for pipes that transport compressed air.

- Red with White Lettering - This red label with white lettering is used for all fire quenching fluids. This could include water, but only if the pipe is used just for fire quenching.

- Orange with Black Lettering - Orange labels with black lettering are for toxic or corrosive fluids. Most acids will need to use this type of pipe marking.

- Brown with White Lettering - This option is for all combustible fluids.

Basic Identification Colours

Water	Green	14-C-53
Steam	Crimson Red	04-D-45
Fire Fighting	Signal Red	04-E-53
Oils (combustible liquids)	Dark Brown	06-C-39
Chemicals (treatment)	Orange	06-E-51
Gases (process and added)	Ochre	08-C-35
Acids & Alkalis	Purple	22-D-45
Air	Light Blue	20-E-51
Process Effluents (drain/vent/flare)	Black	00-E-53

Safety Colours

HAZARD OR SIGN	SAFETY/SEC. COLOUR	BS 4800 CODE REF.
SAFETY	Grass Green	14-C-39
ATTENTION	Golden Yellow	08-E-51
DANGER	Signal Red	04-E-53
MANDATORY	Blue	18-E-53
ALERT	Yellow	10-E-53
ELECTRICAL SERVICE	Orange	06-E-51
TRAFFIC LINES	White	00-E-55

WATER & SANITATION

Water is considered as an chemical in science which contains hydrogen & oxygen. But it is the most essential consumable available on earth sustaining of life. The amount of fresh water on earth is limited and due to misuse, increase in world population its availability and quality is under constant pressure. Preserving the quality of fresh water is important for the drinking-water supply, food production and recreational water use. Water quality can be compromised by the presence of infectious agents, toxic chemicals, and radiological hazards.

It is imporatnt for an Facility/Property Managers to understand the ever increasing water problems and how to handle it. The handling means how best you can avoid wastage, have maximum recycling and conservation of this important natural resource.

Global Water Scenerio

Globally humans started getting a feel of water shortage, the recent example is Cape Town in South Africa where due to severe drought, water was rationed and tourists were asked not to visit. It was estimated that in near future world would be facing huge crisis on waterfront. As per WHO an estimated 2,000 children under the age of five die every day from water borne diseases and 780 million people lack access to safe drinking water. Close to 25% of the earth's population is forced to rely on contaminated water sources for basic needs and people are getting sick from these contaminated water.

Major Source of water are rains, rivers, ponds, dam, under ground water etc.

Access to water solves only a part of the problem. Traditional solutions like bore wells and water delivered by tanker trucks are viable options for water access, but they can easily become compromised. The way the bore wells are used extensively for water supply, ground water level is extensively dropping and if corrective actions are not taken, the days are not far when ground water will be completely dried.

Global Water Consumption

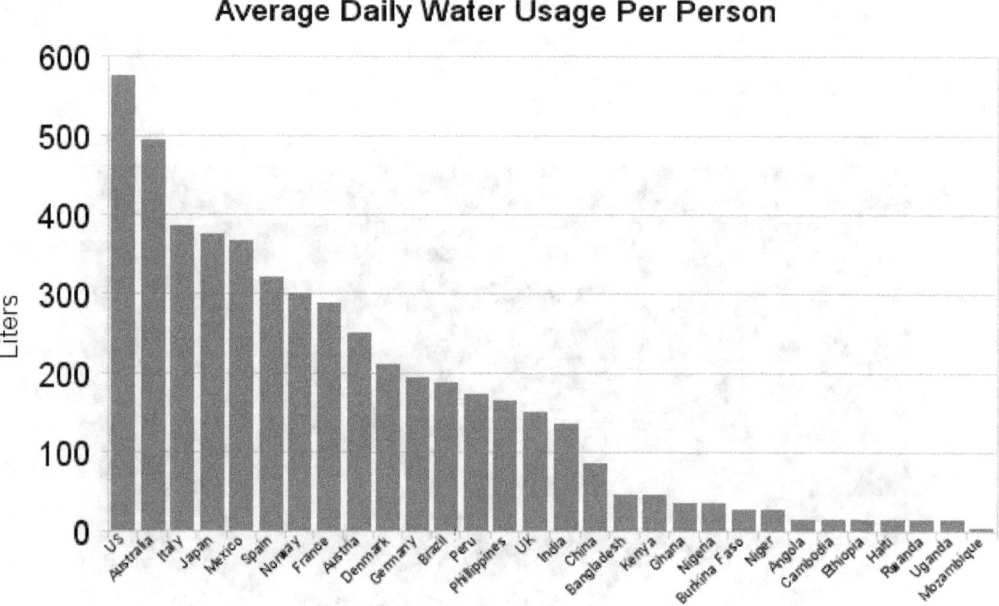

Water Scenario in India

As per the report of UNISEF, despite booming economy, India is struggling with water insecurity and poor water quality and this remains a major cause of child mortality and morbidity, especially among the poor. India lost more than 600,000 children under 5 in 2010 due to WASH (Water, Sanitation and Hygiene) related diseases like diarrhoea and pneumonia.

Water Stress Map

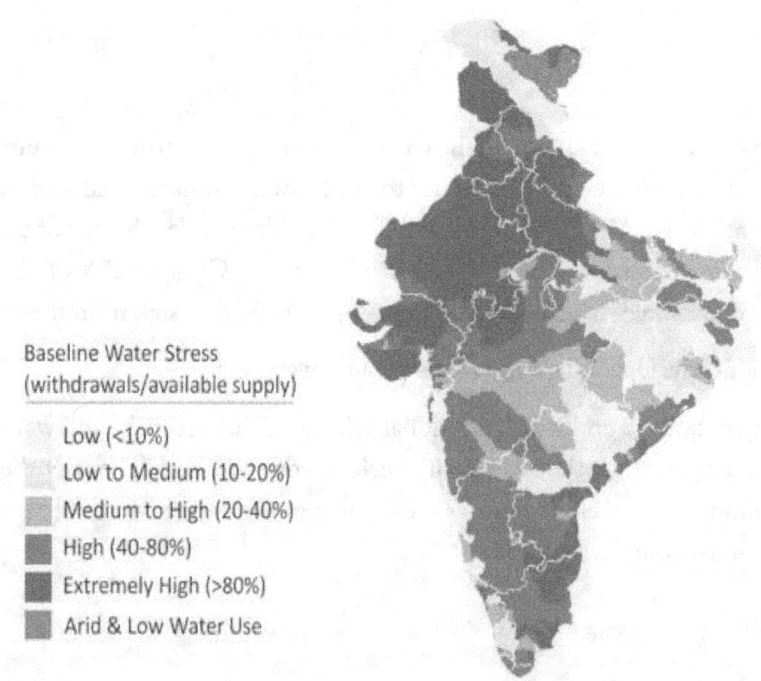

Given increasing water scarcity and the associated deterioration of the quantity and quality of water sources in many parts of the world; many "tools" are developed to map water scarcity risk or water risk. Typically, these tools are based on estimates of the average water supply and demand in each administrative jurisdiction. Often, water risk is basin-wide, while decisions are made at the city or county level. Therefore, the analyses on which such tools are based understate the potential water risk. In most places, even if the resource is not over-appropriated on average, persistent shortages induced by climate variation can lead to stress. A clear understanding of shortages, in terms of the magnitude, duration and recurrence frequency will better inform localities.

Average Domestic Water Consumption: As per the studies undertaken by Mr. Abdul Shabam and RN Sharma and their article published in Economic and Political Weekly, dated June 9, 2017, on an average every Indian house consume 400 liters of water per day and per capita water consumption is 51.51 litres per day.

As per IS:1172-1993,

1. A minimum of 200 liters water per capita per day should be provided for domestic consumption in cities with flushing systems.

2. It also said the amount of water supply can be reduced to 135 litres per capita per day for Low Income Group and weaker section of the society living in small towns.

3. It further depends on Municipals Corporations of individual cities to decide on per capita consumption of water.

   a)   Municipal Corporation of Greater Mumbai consider 135 liter water per capita per day,

   b)   Delhi Development Authority consider 225 Litre water per capita per day for domestic use.

The Indian National Commision of Urbanization (1988) recommended that a per capita water supply of 90 to 100 litres per day is needed to lead a hygenic existenance, and emphasised that this level of water supply must be ensured for all citizens.

On an average Indian person uses 130-150 Liter of water everyday, on regular day.

As per the survey conducted by Mr. Abdul Shabam and RN Sharma and their article publieshed in Economic and Political Weekly, dated June 9, 2017:

Domestic Water Consumption Per Household and Per Capita Per Day (In Liters)

Cities	Per Household		Per Capita	
	Mean	Std Deviation	Mean	Std Deviation
Delhi	377.7	256.8	78	49.9
Mumbai	406.8	158.6	90.4	32.6
Kolkota	443.2	233.6	115.6	64.9
Hydrabad	391.8	172	96.2	43.8
Ahmedabad	410.9	224.1	95	54.6

Activity wise Distribution of Water Consumption in Cities in Percentage

Activity	Delhi	Mumbai	Kolkata	Hydrabad	Ahmedabad
Bathing	31.7	23.7	37.1	25.6	22.8
Cloth Washing	14.2	24.3	14	20.9	21.4
Drinking	5	4.2	2.6	4.3	4.9

(Contd.)

Activity	Delhi	Mumbai	Kolkata	Hydrabad	Ahmedabad
Cooking	3.7	1.7	2.3	3.1	3.3
Toilets	16.5	21.6	15.9	24.1	19.1
House Cleaning	7	6.6	11.7	3.5	12.4
Utensil washing	16.5	17.4	16.1	16.5	15.2
Others	5.6	0.5	0.3	2	0.9
Total	100	100	100	100	100

Global Average Water Use in Commercial Building

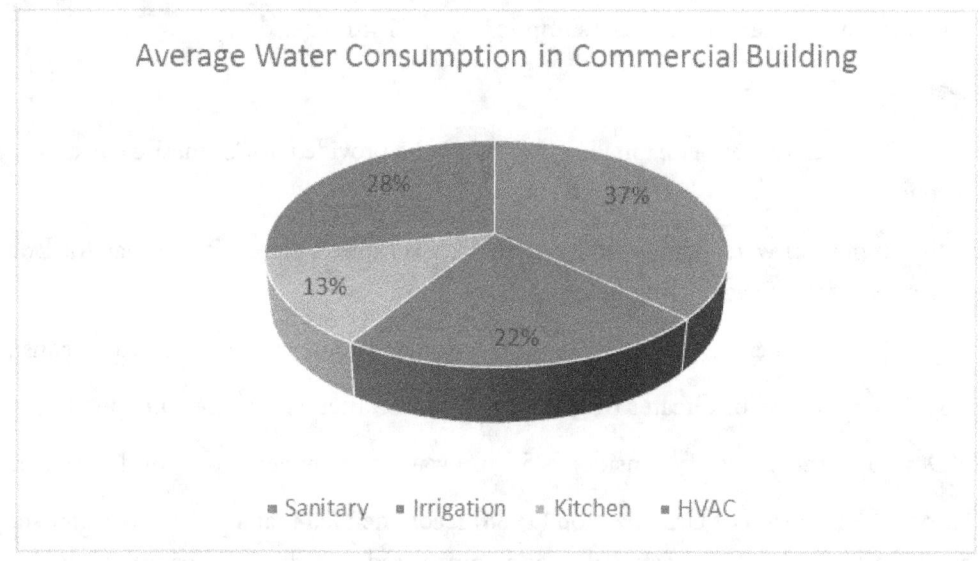

As per IS 1172:1993, clause 4.1 Out of 150 to 200 litres per head per day, 45 litres per head per day may be taken for flushing requirements and the remaining quantity for other domestic purpose.

IS 1172:1993 Domestic water requirement for buildings other than residential is as follows:

Sl. No.	Type of Building	Consumption Per Capita Per Day (Liters)
1	Factories with Bathrooms	45
2	Factories without bathrooms	30
3	Hospital (Including Laundry)	
A	Up to 100 Beds	340
B	Above 100 Beds	450
4	Hostels	135
5	Hotel	180
6	Offices	45
7	Restaurants	70 Per seat
8	Cinema, Concert hall, Theatres	15 per seat
9	Day School	45 per head
10	Boarding School	135 per head
Note: For fire demand in building refer- IS 9668:1981		

Drinking Water Quality

Drinking water may contain many harmful constituents, which can severely affect health of humans and animals. In order to maintain a better quality of drinking water every country has defined parameters.

The World Health Organization has issued Guideline for Drinking Water Quality (GDWQ), include the below mentioned recommended limits on naturally occurring constituents, which may have direct impact on human health:

- Arsenic   10µg/l
- Barium   10µg/l
- Boron   2400µg/l
- Chromium   50µg/l
- Fluoride   1500µg/l
- Selenium   40µg/l
- Uranium   30µg/l

Organic species

- Benzene   10µg/l
- Carbon tetrachloride   4µg/l
- 1,2-Dichlorobenzene   1000µg/l
- 1,4-Dichlorobenzene   300µg/l
- 1,2-Dichloroethane   30µg/l
- 1,2-Dichloroethene   50µg/l
- Dichloromethane   20µg/l
- Di(2-ethylhexyl)phthalate   8 µg/l
- 1,4-Dioxane   50µg/l
- Edetic acid   600µg/l
- Ethylbenzene   300 µg/l
- Hexachlorobutadiene   0.6 µg/l
- Nitrilotriacetic acid   200µg/l
- Pentachlorophenol   9µg/l
- Styrene   20µg/l
- Tetrachloroethene   40µg/l
- Toluene   700µg/l
- Trichloroethene   20µg/l
- Xylene   500µg/l

The BIS drinking water specification (IS 10500:1991) was prepared and approved in 1983 and its most recent revision dates back to July 2010 (Amendment No. 3).

The standard was adopted by the Bureau of Indian Standards with the following objectives -

- To assess the quality of water resources, and
- To check the effectiveness of water treatment and supply by the concerned authorities.

They apply to drinking water supplied by different Authorities/ Agencies/ Departments of State Governments and Central Government, wherever applicable in the country. They also apply to water supplied by Non Government or Private Agencies for human consumption in any place of the country.

The Drinking Water Standard in India is guided by IS 10500 : 1991 and is as appended below:

- Turbidity NTU Max.   5
- pH Value   6.5 to 8.5
- Hardness (CaCO3) mg/L   300, in absence of alternate source Max. permitted is 600
- Iron (Fe) mg/L   0.3, in absence of alternate source Max. permitted is 1
- Chloride (Cl) mg/L   250, in absence of alternate source Max. permitted is 1000
- Residual free Chloride, mg/L   0.2

Desirable Characteristic

1	TDS mg/L	500, in absence of alternate source Max. Permitted is 2000
2	Calcium (Ca) mg/L	75, in absence of alternate source Max. Permitted is 200
3	Copper (Cu) mg/L	0.05, in absence of alternate source Max. Permitted is 1.5
4	Manganese (Ma) mg/L	0.1, in absence of alternate source Max. Permitted is 0.3
5	Sulphate (So4) mg/L	200, in absence of alternate source Max. Permitted is 400
6	Nitrate (No2) mg/L	45, in absence of alternate source Max. Permitted is 100
7	Fluoride (F) mg/L	1.0, in absence of alternate source Max. Permitted is 1.5
8	Phenolic Compound (C4H5OH)	0.001, in absence of alternate source Max. Permitted is 0.002 mg/L
9	Mercury (Hg) mg/L	0.001, No Relaxation
10	Cadmium (Cd) mg/L	0.01, No Relaxation
11	Selenium (Se) mg/L	0.01, No Relaxation
12	Arsenic (As) mg/L	0.05, No Relaxation
13	Cyanide (Cn) mg/L	0.05, No Relaxation
14	Lead (Ph) mg/L	0.05, No Relaxation

15	Zinc (Zn) mg/L	05, in absence of alternate source Max. Permitted is 15
16	Amonic Detergent	2, in absence of alternate source Max. Permitted is 10 (MBAS) mg/L
17	Chromium (Cr8+) mg/L	0.05, No Relaxation

IS 10500:1991, Table I, Test Characteristic for Drinking Water

S.N	Characteristic	Desirable Limit	Permissible Limit	Undesirable Effect post Desirable effect limit
1	Polynuclear Aromatic Hydrocarbons (PAH)g/L	0	0	May be carcinogenic
2	Mineral Oil mg/L	0.01	0.05	Undesirable taste & Oder
3	Pesticides mg/L	Absent	0.001	-
4	Radioactive Material			
	Alpha emitters Ba/L	-	0.1	-
	Beta emitters pci/L	-	1	-
5	Alkalinity mg/L	200	600	Taste becomes unpleasant
6	Aluminium AL, mg/L	0.03	0.2	Cause dementia (Brain related disease)
7	Boron mg/L	1	5	-

Sanitation

The word Sanitation refers to the provision of sufficient facilities and services for the safe disposal of human waste. For a neat and clean environment, one needs sufficient number of facilities, which can cater for the available population. In a situation where you do not have sufficient number of urinals and water closet (toilet), you will not be able to maintain a clean environment. The word 'sanitation' also refers to the maintenance of hygienic conditions, through services such as garbage collection and wastewater disposal.

The main purposes of sanitation are to provide a healthy living environment for everyone, to protect the natural resources (such as surface water, groundwater, soil), and to provide safety, security and dignity for people when they defecate or urinate. Humans rights with respect to water & sanitation was recognized and adopted by the United Nations (UN) general assembly in 2010. It is also recognized by international law through treaties, declarations and other standards.

In view of the importance of sanitation the BIS has defined standards for minimum requirement of sanitation for public places, factories, offices and other places where humans are gathered in large numbers.

As per IS 1172:1993 clause 6.4.1 the minimum sanitary convenience required at any railway station, bus stop or bus terminal & sea ports is as appended below:

Nature of Station	WC for Males	WC for Females	Urinals of Males only
Junction stations & intermediate stations & bus stations	3 for first 1000 and 1 for every additional 1000	4 for first 1000 and 1 for every additional 1000	4 for first 1000 and 1 for every additional 1000
Terminal stations & Bus terminals	4 for first 1000 and 1 for every additional 1000	5 for first 1000 and 1 for every additional 2000	6 for first 1000 and 1 for every additional 1000

As per IS 1172: 1993 clause 6.4.1, the sanitary conveniences required at airports:

Type of Airport-Domestic	WC for Males	WC for Females	Urinals of Males only
For 200 persons	5	8	6
For 400 persons	9	15	12
For 600 persons	12	20	16
For 800 persons	16	26	20
For 1000 persons	18	29	22
International Airport			
For 200 persons	6	10	8
For 600 persons	12	20	16
For 1000 persons	18	29	22

Note: Separate provisions should be provided for staff and workers. At least one Indian style closet (toilet) to be provided in each toilet. Assume 60 male to 40 female.

As per IS 1172: 1993 clause 6.4.1, the sanitary conveniences required for office buildings:

S. No.	Fitments	Male	Female
1	Water Closets	1 for every 25 person	1 for every 15 person
2	Ablution Taps	1 in each water closet	1 in each water closet
		One water tap with drain arrangement shall be provided for every 50 persons or part thereof in vicinity of water closet & urinals	
3	Urinals	Nil up to 6 persons	
		1 for 7 to 20 persons	
		2 for 21 to 45 persons	
		3 for 46 to 70 persons	
		4 for 71 to 100 persons	
		From 101 to 200 persons add at rate of 3%	
		For over 200 persons add at rate of 2.5%	
4	Wash Basins	01 for every 25 person	
5	Drinking water fountains	For every 100 persons with a minimum of one on each floor	
6	Cleaners Sink	1 per floor, preferably adjacent to sanitary room	

As per IS 1172: 1993, table 03, clause 5.3, the sanitary conveniences required for factories:

Sr. No	Fitments	Male	Female
1	Water Closet	1 for 1 to 15	1 for 1 to 12
		2 for 16 to 35	2 for 13 to 25
		3 for 36 to 65	3 for 26 to 40
		4 for 66 to 100	4 for 41 to 57
			5 for 58 to 77
			6 for 78 to 100
		From 101 to 200 add @ 3%	From 101 to 200 add @ 5%

		For over 200 add @ 2.5%	For over 200 add @ 4%
2	Ablution Taps	1 in each water closet	1 in each water closet
		01 water tap with drainage arrangement to be provided for every 50 person or part there of near water closet & urinals.	
3	Urinals	Nil up to 06 persons	
		1 for 7 to 20 persons	
		2 for 21 to 45 persons	
		3 for 46 to 70 persons	
		4 for 71 to 100 persons	
		101 to 200 persons add @ 3%	
		For 200 persons & above add @ 2.5%	
4	Washing tap with drainage arrangement	1 for every 25 persons or part there of	
5	Drinking water fountain	01 for every 100 person or part thereof with a minimum of one on each floor	
6	Bath preferably shower	As required for particular trade or occupations.	

Note:

1. For many trade of dirty & dangerous character, law requires more provisions that are extensive.

2. Crèches, where provided, shall be fitted with water closet (one for 10 person or part thereof) and wash basins (one for 15 person or part thereof) and drinking water tap with drainage arrangements (one for 50 person or part thereof)

As per IS 1172: 1993, table 04, clause 5.3, the sanitary conveniences required for Cinemas, Concert Halls & Theatres:

Sr. No	Fitments	For Public -Male	For Public -Female	For Staff- Male	For Staff- Female
1	Water Closet	1 per 100 to 400 person, at add @ 1% per 250 person thereof	3 per 100 to 200 person, for over 200 person at @ 2% per 100 person thereof	1 for 1 to 15 person, 2 for 16 to 35 persons	1 for 1 to 12 person, 2 for 13 to 25 persons
2	Ablution taps	1 in each water closet	1 in each water closet	1 in each water closet	1 in each water closet
3	Urinals	1 for 25 persons or part there of	-	Nil up to 06 persons 1 for 7 to 20 person 2 for 21 to 45 person	
4	Wash basins	1 for every 200 person or part there of	1 for every 200 person or part there of	1 for 1 to 15 person 2 for 16 to 35 person	1 for 1 to 12 person 2 for 13 to 25 person of
5	Drinking water	1 for every 100 person or part thereof			

Note:

1. Some of the water closet can of European style, if desired.

2. It may be assumed that two third of persons may be male and one-third persons may be female.

3. Provision of water tap may be also be made in public place of drinking water fountains, the scale of which may be 1 per 100 person or part thereof

You may also refer IS4878-1986 The Bye laws for construction of cinema halls.

As per IS 1172: 1993, table 06, clause 5.3, the sanitary conveniences required for Cinemas, Concert Halls & Theatres:

Sl No	Fitments	Requirements
	Indoor Patient wards (For male & females)	
1	Water Closet	1 for every 8 beds or part thereof
2	Ablution taps	1 in each water closet, plus one water tap with draining arrangements in the vicinity of water-closets & urnials for every 50 beds or part thereof
3	Wash basins	2 up to 30 beds; add 1 for every 8 beds or part thereof
4	Baths	1 bath shower for every 8 beds or part thereof
5	Bed pan washing sinks	1 for each ward
6	Cleaner's sink	1 for each ward
7	Kitchen sinks & dish washers (where kitchen is provided)	1 for each ward

IS 1172: 1993, table 06, clause 5.3, the sanitary conveniences required for Hospitals:

Sl No	Fitments	Requirements Male	Requirements Female
	Outdoor Patient wards & visitors (For male & females)		
1	Water Closet	1 for every 100 persons	2 for every 100 persons
2	Ablution taps	1 in each water closet	
	1 water tap with drainage arrangement to be provided for every 50 person or part thereof in the vicinity of water closet or urinals		
3	Wash basins	1 for every 100 persons	1 for every 100 persons
4	Urinals	1 for every 50 persons or part thereof	-
4	Drinking water fountains	1 for every 500 persons or part thereof	

IS 1172: 1993, table 10, clause 5.3, the sanitary conveniences required for Education Institutions:

Sr No.	Fitments	Nursery Schools	Education Institutions (Non Residential)		Education Institutions (Non Residential)	
			Boys	Girls	Boys	Girls
1	Water Closets	1 per 15 & part there of	1 per 40 & part there of	1 per 25 & part there of	1 per 8 & part there of	1 per 6 & part there of
2	Ablution taps	1 in each water closet	1 in each water closet	1 in each water closet	1 in each water closet	1 in each water closet
		1 water tap with draining arrangement shall be provided for every 50 pupils or part thereof in the vicinity of water closets and urinals				
3	Urinals	-	1 per 20 pupils or part thereof	-	1 per 25 pupils or part thereof	-
4	Wash basins	1 per 15 pupils or part thereof	1 per 60, Min 2	1 per 40, min 2	1 per 8 pupils or part thereof	1 per 6 pupils or part thereof

5	Baths	1 bath sink for every 40 pupils or part thereof	-	-	1 per 8 pupils or part thereof	1 per 6 pupils or part thereof
6	Drinking water fountain	1 per 50 pupils or part thereof	1 per 50 pupils or part thereof	1 per 50 pupils or part thereof	1 per 50 pupils or part thereof	1 per 50 pupils or part thereof
7	Cleaners Sink	Minimum 1 per floor				

Note: For teaching, staff the schedule of fitments to be provided as in the case of office building.

You can also refer:

IS 2064:1993   Code for practice for selection, installation and maintenance pf sanitary appliances (second revision)

4878:1986   Byelaws for construction of cinema buildings (first revision)

9668:1990   Code of practice for provisions and maintenance for water supplies for fire fighting

BUDGETING & BUDGETARY CONTROL

What is a BUDGET?

The word Budget is derived from the French word 'Budgette' which means small leather bag.

Budget is prepared considering revenue and expenses expected in upcoming financial year. A budget is defined as a comprehensive and coordinated plan expressed in financial terms for the resources and operations of an enterprise for some specified period.

What is a budget?

➢ Budget means a Plan for upcoming year

➢ It's a cap on expenses

➢ It's a Schedule or a plan for revenue & expenses

➢ It's a Reality Check for income & expenditure

➢ It's an allocation of funds

Elements of a Budget

There are two important elements of budget:

1. **Personal**

   Persons are very important factors for creating or preparing the budget. To ensure a best and accurate budget organization shall ensure that people works on budget with utmost sincerity and accountability. Budget shouldn't be prepared with a casual approach. The organization should provide enough time to the concern employee to prepare budget as far as accurate.

2. **Process**

   The persons responsible for preparing budget shall have access to all the required information's. It is also suggested that if a new person is deputed to prepare a budget you should ensure that the person is briefed enough and should be updated with your company's process and required assumptions, as the process to prepare a budget differs from company to company.

   One should also brief the concern person with companies plan for the year, so that the person preparing a budget has a vision. Budget should be ready well before the year start; this will help in getting timely approvals from the management and helps in planning business in advance.

   It's always better you prepare a budget for a year and monitor the same on monthly basis and check for variance. Every company has their own standard for defining variance percentage allowed. As a responsible person one shall always look for reasons behind variance and try and control them. Apart from monthly monitoring of budget expenses one should also monitor the same on quarterly basis, as the quarterly budget control will be another important factor.

Factors Affecting Budget

1. Revenue: Budget will get badly affected if the projected revenue or income is affected and is below the expected revenue.

2. Expenditure: In case any unplanned expense due to requirement of major expenses such as replacement of any equipment because of failure.

3. Market Condition: In case of sudden collapse of market, the business gets severely affected and affects the revenue. In this case the budget goes for a toss.

4. Government Policy: Changes in government policies such as taxation, or reversal of rebate given on taxes etc. etc. can affect the budget.

5. Internal Factors: Promotional program, Manufacturing Processes etc. Change in business strategy, new business

Budgeting not only suggests what should happen but should also make things happen.

OPERATIONS AND RESOURCES

The budget is prepared in two important headings:

a) Opex: Operating Budget

b) Capex: Capital Budget

The operations are reflected in revenues and expenses. A budget should quantify the revenue, operating expense and capital expense in INR (or the concern countries' currencies).

The planning of Capital budget means the planning of the various assets and the sources of capital to finance these assets.

FINANCIAL TERMS

Budgets are prepared in financial terms that are in terms of monetary value such as:

a) Assets:

Assets are a resource of money value such as stocks, bonds, real estate and cash. Assets can be fixed and moveable assets.

b) Capital:

Capital is a kind of fund which is required to spend on major expenses, which are not routine in nature. E.g. setting up of new office, purchase of equipment & its installation.

c) Depreciation

Once the capital is spent an amount from balance sheet is being deducted as depreciation which you can consider as returning of fund used to the source and also as per taxation policy you can get rebate on tax on depreciation value. Depreciation is also known as decline in valuation of investment.

d) Capital Appreciation:

If a fixed asset such as land or building is purchased by the company and the appreciation in its value due to market conditions such appreciation is known as Capital Appreciation.

e) Currency Risk:

Valuation of Currency fluctuation in global market will always affect the rupee value of an investment and company's revenue.

f) Net profit margin:

A measure of a company's profitability and efficiency calculated by dividing a measure of net profits (operating profit minus depreciation and income taxes) by sales.

g) Tax Deduction at Source (TDS):

Percentage of tax deducted from the income by a company and paid to the government as tax on your behalf.

h) FBOI (Fully Burdened Operating Income):

The income left out after deduction of all expense, taxes, depreciation from the total revenue.

SPECIFIED FUTURE PERIOD

A budget is related to a specific period of time of one year, otherwise it is meaningless. In case of Indian companies the budget period starts from 1st April and ends on 31st March. In case of US companies the budget starts on 1st January and ends at 31st December.

In similar way it depends on financial year of the countries from which the company belongs to.

COMPREHENSIVENESS

A budget has to be comprehensive and should contain all the aspects of revenue and expense. All the activities and operations of a project should be included in the budget. A comprehensive budget increases chances of accountability and transparency.

COORDINATION

A budget has to be prepared in coordination and consultation with the entire stake holder. The stakeholder can be the senior most person of the organization and can also be the person at lower management level. Budgets are prepared for different segments/divisions/ facets/ activities of an organization in harmony with each other.

Budget Involves

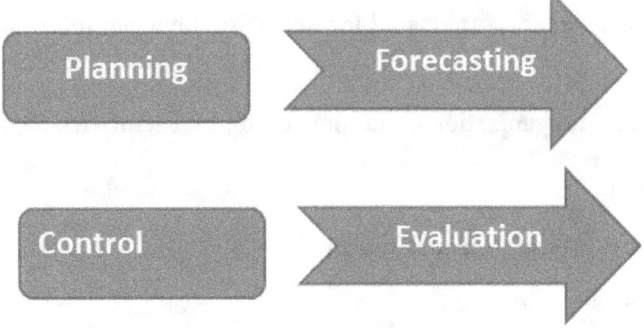

Why use a budget?

In order to ensure doing a profitable business at any scale needs better planning and timely review of company's financials. Budget is prepared to ensure:

- Explicit statement of expectations on revenue and expenses.
- Communication among all the stake holders
- Coordination among all the stake holders
- Prioritize Wants, Organize Needs, within the realm of what we Can

Why budget is important for business?

Expenditure Control

Budgeting helps in capping the expenditures on various fronts. Budget is prepared basically to ensure that the funds are not wasted on unnecessary items or works or unnecessary capital expenses.

Creates Financial Roadmaps

A budget provides financial roadmap for smooth operations of a business. Companies normally consider previous year's budget as benchmark and compare variances accordingly for the present year. The increase in variances of expenditure can be due to a good growth of business. In this case you can ask for increase in budget expenditure from Management. If there is increase in variances on expenditure without any growth of business then in that case you will have to put a hold on unnecessary expenses.

Future Growth Plan

Budgets can be constructively used to plan for future business growth and expansion. Money saved from day today business expenditures can be saved as special reserve account to be used for new business opportunities. The same saved amount can also be used for paying regular business expenses at the time of slowdown in economy.

Tools

You can use one out of many accounting software available in the market. This software helps you with better records of past expenses and revenue generations, it helps a corporates to automate their budget process and easily track their expenses and track them electronically. This software also helps in easily accessing all the information's required for preparing the budget.

These software packages are a very important tool for managing financial information and reviewing information in a real-time format.

Procedure in budget preparation organization & administration by Facility Management professionals

- Fixation of budget period.
- Check with business units for their expansion plan
- Check the past year expense trend

- Discuss with your team on expenses under various head

- Jot down the details of expenses under various head

- Determine the key factors

How can you make budget effective

- The Facility Manager shall ensure that the forecasting is accurate

- The budget should be based on organisational goals

- The Information should be prepared in time and should be accurate

- Prepare the budget with inputs collected from multilevel

- As an Facility Manager one should review the budget regularly

How can you prioritize your budget?

1. Eliminate minor but needless costs: One should start looking for minor spends without which you can easily manage, which will help you with small cost eliminations. For example, if you have a plumber and electrician in your facility, instead you can hire multi skilled technician who can perform both the task in a cost of slightly more than one technician.

   Another example can be replacing tissue paper dispenser with hand drier. This will eliminate recurring cost on buying of tissue papers.

2. Reduce larger expenses: These recommendations are decidedly more painful & stressful. You can consider reducing your electricity bill by taking measures to reduce and control your monthly electricity bills. You can also initiate action to reduce on transportation cost by consolidations.

   On the assumption that those kinds of changes may be too wrenching, here are some other specific areas where many people can find savings:

How to Control Budget

➢ It is the responsibility of the person heading Facility Management division to ensure that he/she keeps an eye on regular expense and compare it with the planned budget at every month end.

➢ He /She should cross check for any new expenses brought to him/she for approvals. And ask number of questions and shouldn't approve until he/she gets completely satisfies.

➢ One should also try and renegotiate with vendors on fixed costs and try and bring the cost down.

➢ One should keep a close watch on all the expenses. And should even challenge the budgeted expense if needed.

   In case of increase in variance the responsible person should put a strict control in place to control the expense.

Draft chart for Budgeting (for reference only)

	Previous year Expense month wise			Budget for next financial year month wise		
	Jan	Feb	March.........	Jan	Feb	March......
Salaries & Wages						
Burdens -Paid Absences						
Burdens - Group Insurance (L,H & D)						
Burdens - Retirement Plan Pension						
Payroll Burdens – Other						
Employee Bonuses & Commissions						
Incentive Compensation - Amortization						
Contract Labor						
Contract Personnel - Labor						
Housing & Living Costs - Accommodations						
Housing & Living Costs - Cost of Living (Food)						
Conventions, Seminars & Books						
Insurance - Workers Compensation						
Relocation Exp Employees - Taxable						
Medical – Supplies						
Medical – Exams						
Entertainment						
Employee Meals - Business Meetings						
Travel – Transportation						
Travel – Accommodations						
Travel - Employee Meals - Reimbursable						
Gain Loss on Disposal of Assets – Equipment						
Equipment Operating Expenses – Supplies						
Operating Equip. Rented - Short term						
Operating Equip. Rented - Long term						
Equipment Operating Expenses - Tires & Tubes						
Equipment Repair Costs - Heavy Duty Equipment						
Equipment Repair Costs - Service Contracts						

Equipment Repair Costs - Light Duty Vehicles						
Equipment Maintenance						
Equipment Operating Expenses - Fuels						
Equipment Operating Expenses - Oils & Lubricants						
Facility costs - Rental Expense						
Facility Costs - Outside Guard Service						
Facility Costs – Housekeeping						
Facility Costs – Maintenance						
Facility Costs - Moving Expense						
Facility Costs - Temporary Facilities						
Facility Closure Expense-Restructuring						
Heat, Light, Water, Sewer & Power						
Provisions Commissary						
Depreciation						
Supplies						
Small Tools						
Modifications & Rework						
Inspection, Testing and Analysis						
Certification Costs						
Safety Supplies						
Office Supplies						
Printer supplies						
Reprographic Services						
Demurrage						
Freight Expense- Auto Posting						
Land Freight						
Air Freight						
Telephone						
Communications - Cellular Phones						
Computer Software						
Computer Equipment - Maintenance						
Software Lic						
PC Device Cost						
PC support						
IT Service Charge						
Outside Consultants – Meals						

(*Contd.*)

	Previous year Expense month wise			Budget for next financial year month wise		
	Jan	Feb	March.........	Jan	Feb	March......
Consultants – Accommodations						
Consultants – Transportation						
Legal Services Expenses						
Auditing Services						
Tax Services Expense						
Facility Insurance						
Insurance - Excess Liability						
Insurance - General Liability						
Insurance - Auto Liability						
Insurance - Contractors Equipment						
Courier Services						
Subscription for magazine/books						
Waste Management						

LABOR LAWS & COMPLIANCE

Various Acts

1. The Factories Act
2. The Contract Labour (R&A) Act
3. The Payment of Wages Act
4. The Minimum Wages Act
5. The Payment of Bonus Act
6. The Employment Exchanges Act
7. The Provident Fund Act
8. The Profession Tax Act
9. The Work men Compensation Act
10. The Maternity Benefit Act
11. The Industrial Employment (Standing Orders) Act
12. The Bombay Labour Welfare Fund Act
13. The Payment of Gratuity Act
14. The Bombay Shops and Establishment Act
15. The Industrial Disputes Act
16. The Water (Prevention & Control of Pollution) Act
17. The Maharashtra Recognition of Trade Union (MRTU) and Prevention of Unfair Labour Practices (PULP) Act
18. The Employees' State Insurance Act

Applicability of Acts

The Factories Act

Applicability : Any premises including precincts thereof where ten or more workers are working or were working on any day of preceding twelve months and in any part of which a manufacturing process is being carried on with the aid of power, where twenty or more workers are working or were working on any day of preceding twelve months and in any part of which a manufacturing process is being carried on without the aid of power.

Scope: An Act to consolidate and amend the law regulating labor in factories

Comments: This Act covers basic rules and regulations for starting of a new factory and maintaining it consistently under the conditions prescribed in the Act for safety and health and welfare of the employees.

The Payment of Wages Act

Applicability: Every Factory

Scope: An Act to regulate the payment of wages to certain classes of employed persons.

Comments: This Act covers various rules regarding payment of wages like date of payment, place of payment, mode of payment, verification by principle employer and records thereof. It also describes eligible deduction from wages and related procedures.

The Minimum Wages Act

Applicability: Every Factory or Establishment

Scope: An Act to provide for fixing minimum rates of wages in certain employments.

Comments: This Act covers rules & regulations regarding minimum wages applicable to nature of industry which is decided based on process carried out. It also covers the index for Dearness Allowance and HRA rules.

The Contract Labour (Regulation & Abolition) Act

Applicability: Every establishment in which twenty or more workmen are employed or were employed on any day of the preceding twelve months as contract labor through contractor for the work to be carried out which is not directly connected to manufacturing activities, to every contractor who employs or who employed on any day of the preceding twelve months twenty or more workmen.

Scope: An Act to regulate the employment of contract labor in certain establishments and to provide for its abolition in certain circumstances and for matters connected therewith.

Comments: The principle employer and the contractor should obtain permission by applying for registration to The Competent Authority.

The Payment of Bonus Act

Applicability: Every factory & every other establishment in which twenty or more persons are employed on any day during an accounting year.

Scope: An Act to provide for the payment of bonus to persons employed in certain establishments on the basis of profits or on the basis of production or productivity and for matters connected therewith.

Comments: Every factory or establishment completed 5 years or the year in which it earns profits whichever is earlier, this Act is applicable then onwards. Registers to be maintained and Returns to be submitted on applicability of this Act.

The Employment Exchanges (Compulsory Notification of Vacancies) Act

Applicability: Every establishment employing twenty five or more persons.

Scope: An Act to provide for compulsory notification of vacancies to Employment Exchanges.

Comments: Needs to comply by notifying vacancies every quarter and to submit the returns quarterly as well as biannually.

The Provident Fund Act

Applicability: Every establishment which is a factory specified in Schedule 'I' and where twenty or more persons are employed.

Scope: An Act to provide for the institution of Provident Funds for employees in factories and other establishments

Comments: Registration under the Act is mandatory for all. Principle employer is responsible for deductions and remittance of PF contributions under this Act even for contract employees also. Monthly, yearly returns are required to be submitted in the prescribed format.

The Profession Tax Act

Applicability: Every Factory/Establishment

Comments: Registration under the Act is mandatory for all. Principle employer is responsible for deductions and remittance of Profession Tax under this Act even for contract employees also. Monthly, yearly returns are required to be submitted in the prescribed format.

The Workmen's Compensation Act

Applicability: Every Factory

Scope: An Act to provide for the payment by certain classes of employers to their workmen of compensation for injury by accident.

Comments: The Act speaks about the compensation payable in case of disablement due to occupational hazards. ESI can be substituted by having insurance policy under the Act to take care of compensation payable, if any.

The Maternity Benefit Act

Applicability: Every shop & establishment in which ten or more persons are employed on any day of the preceding twelve months or where Employees State Insurance Act is applicable for the time being.

 Scope: An Act to regulate employment of women in certain establishments for certain period before and after child-birth and to provide for maternity benefits and certain other benefits.

Comments: This act describes maternity benefits to the working women in certain establishments.

The Industrial Employment (Standing Orders) Act

Applicability: Every industrial establishment wherein fifty or more workmen are employed on any day of preceding twelve months.

Scope: An Act to provide for defining with sufficient precision certain conditions of employment in industrial establishments.

Comments: This Act regulates basic terms and conditions of employment. The employer can get its rules and regulations certified from the competent authority or The Model Standing Orders described in the Act gets applied to the establishment.

The Bombay Labor Welfare Fund Act

Applicability: Applies to every factory within the meaning of Factories Act, 1948, every establishment within the meaning of Shop & Establishment Act, 1948, branches/ Depts. situated in the same premises or different places thereof.

Scope: An Act to provide Constitution of a fund for the financing of activities to promote welfare of labor in the state of Maharashtra

Comments: The employer has to deduct the amount specified every six month from the eligible employee's salary/wages and remit the same to the Welfare Fund. The amount collected is used for the welfare of labors by various means.

The Payment of Gratuity Act

Applicability: Every factory, every shop & establishment in which ten or more persons are employed, or were employed on any day of the preceding twelve months.

Scope: An Act to provide for a scheme for the payment of gratuity to employees engaged in factories, mines, oilfields, plantations, ports, railway companies, shops and other establishments and for matters connected therewith or incidental thereto.

Comments: Every employee who has completed his 4 years and 240 days' uninterrupted continuous service is eligible to get benefits under this Act. While calculating the refund to the employee the basic pay is to be divided by 26 days and the employee gets 15 days of basic pay for every completed year.

The Bombay Shops and Establishment Act

Applicability: Every Shop and Establishment which is not a Factory under The Factories Act.

Scope: An Act to consolidate and amend the law relating to the regulation of conditions of work and employment in shops, commercial establishments, residential hotels, restaurants, eating houses, theatres and other places, public amusement or entertainment and other establishments.

Comments: This Act covers every shop and establishments which is not a factory and regulates the conditions of employment. The employer has to obtain a license from the authority under the Act and renew it regularly before the expiry.

The Industrial Disputes Act

Applicability: Every factory

Scope: An Act to make provisions for the investigation and settlement of industrial disputes and for certain other purpose

Comments: The act describes the meaning, legality and rules regarding Strike, Lock-out, Lay-off and Retrenchment. It also describes the procedure of making complaints to the competent authority and the settlement thereof.

The Water (Prevention & Control of Pollution) Act

Applicability: Every factory

Scope: An Act to provide for prevention and control of water pollution and the maintaining or restoring of wholesomeness of water, for the establishment, with a view to carrying out of the purposes aforesaid, of Boards for the prevention and Control of Water Pollution, for conferring on and assigning to such Boards powers and functions relating thereto and for matters connected therewith.

Comments: This act covers all rules and regulations related to prevention of Water and Air pollution and control thereto. Every employer has to obtain a license to establish/operate from the MPCM Board.

The Employees' State Insurance Act

Applicability: Every factory & establishment

(for employees drawing wages less than Rs. 15,000/-)

Scope: An Act to provide for certain benefits to employees in case of sickness, maternity and employment injury and to make provision for certain other matters in relation thereto.

Comments: This act covers all rules and regulations related to employee safety & health issues.

The Factories Act

Returns/Registers	Form No.	When to comply
Approval of Plans	I	Before situating a factory or construction or extension of it
Certificate of Stability	1-A	Once every 5 years
Application for Registration	2	After commencement of Factory
Grant and renewal of License	3	Starting and renewal on or before 31st Oct., every year

Returns/Registers	Form No.	When to comply
Notice of Change of Manager	5	As and when Factory Manager changes
Record of white washing, varnishing, painting, etc. & repainting, re-varnishing	8	To be maintained always
Register of workers attending to machinery	10	To be maintained always
Report of examination of lifting machinery/ropes/tackles by competent person	12	Certificate to be obtained annually
Examination of pressure plant by competent person	13	Every Six Months
Externally		Once in Twelve months
Internally		Once in four years
Hydraulic Test		

(*Contd.*)

Returns/Registers	Form No.	When to comply
Register of Compensatory Holidays	14	To be maintained always
Notice of period of work for adults	16	To display and maintain
Register of adult workers	17	To be maintained always
Leave with wages register	20	To be maintained always
Leave Book	21	To be maintained always
Report of accident by the Manager	24	To be submitted to Factory Inspection office as and when accident takes place
Notice of Dangerous Occurrence	24A	Within 12 hours of taking place of such accident
Abstract under the Act	26	To be displayed always
Annual Return	27	To be submitted to Factory Inspection office on or before 1st Feb. every year
Muster Roll	29	To be maintained if Form 17 & 19 are not maintained
Register of accidents & dangerous occurrences	30	To be maintained always
Inspection Book	31	To be maintained always
Notification of Paid Holidays		To display and to submit to Factory Inspection Office every year in January

The Contract Labor (Regulation & Abolition) Act

Returns/Registers	Form No.	When to comply
Application for registration of establishments	I	Every year before 31st December
Certificate of Registration	II	To preserve
Application for license	IV	To be made by contractor
Certificate from Principle Employer	V	Required for submission of application for license
Renewal of License	VI	To be made by contractor every year before 31st October
Register of contractor	VII	To be maintained by Principle Employer
Register of persons employed	IX	To be maintained by the contractor
Muster Roll	XII	To be maintained by the contractor
Register of Wages	XIII	To be maintained by the contractor
Annual Return	XXI	By 15th February every year

The Payment of Wages Act

Returns/Registers	Form No.	When to comply
Abstract under the Act		To display always
Register of Wages	II	To maintain always

Register of deductions for damage or loss	III	To maintain always
Notice of dates of payment		To display always
Register of advances	IV	To maintain always
Annual Return	V	To submit before 15th February every year
Display of rates of Wages		To display always
Inspection Book		To maintain always
All registers and records to be preserved for a period of THREE years.		

The Minimum Wages Act

Returns/Registers	Form No.	When to comply
Muster Roll cum Wage register	II	To maintain always
(Unless registers in Form 17 & 19 appended to Maharashtra Factories Rules, 1963 & a register in Form II appended to the Maharashtra Payment of wages Rules, 1963 are maintained.)		
Inspection Book		To maintain always
Abstract	I	To display always
Annual Return	3	To submit every year before 15th February
Attendance cum Wage slip or Wage Card		To maintain always

The Payment of Bonus Act

Returns/Registers	Form No.	When to comply
Allocable Surplus register	A	To maintain
Set on and Set Off Register	B	To maintain
Payment of Bonus register	C	To maintain
Annual Return	D	To submit every year before 30 days from the date of payment of bonus
Inspection Book		To maintain
Unclaimed Bonus register		To maintain

The Employment Exchanges (Compulsory Notification of Vacancies) Act

Returns/Registers	Form No.	When to comply
Notification of Vacancies		15 days before interview/test
Local Employment Exchange		60 days before interview/test
Central Employment Exchange		
Quarterly Return	ER-I	Within 30 days from the completion of quarter.
Biannual Return (Occupational Return)	ER-II	Once in a two years on a date to be specified in the official Gazette

The Provident Fund Act

Returns/Registers	Form No.	When to comply
Declaration and Nomination	2	On becoming the member of EPF scheme
Contribution Card	3-A	Yearly submission before 30th April
Return of employee qualifying to become member of the fund for the first time	5	To submit before 15th of every month
Return of Ownership to be sent to the Regional State Commissioner	5-A	Once
Consolidated annual contribution statement	6-A	Yearly submission before 30th April
Cancellation/Change of Nomination	8	As and when there is change in the nomination of the employee
Return of members leaving service	10	To submit before 15th of every month
Statement of contribution for the month	12-A	To submit before 25th of every month
Transfer of EPF account	13	As and when employee joins other company
Application of withdrawal of EPF	19	On leaving of service of a member
Challan (No. 1, 2, 10, 21, 22)		To submit before 15th of every month
Inspection Book		To maintain always

The Workmen's Compensation Act

Returns/Registers	Form No.	When to comply
Statement of depositing compensation	A	At the time of depositing the compensation in respect to death
Application by dependents for deposit of compensation	G	To be made by dependents
Repot of fatal accidents	EE	At the time of occurrence of fatal accident

The Maternity Benefit Act

Returns/Registers	Form No.	When to comply
Abstract		To display always
Annual Return	11	To submit before 15th January every year

The Industrial Employment (Standing Orders) Act

Returns/Registers	Form No.	When to comply
Register of Standing Orders or Model Standing Orders	E	To maintain always

Submission of draft amendments	A	As and when required
Notice of discontinuance/restart of a shift	IV-A	As and when required

The Bombay Labor Welfare Fund Act

Returns/Registers	Form No.	When to comply
Register of Wages	B	To maintain always
(Unless registers in Form 17 & 19 appended to Maharashtra Factories Rules, 1963 & a register in Form II appended to the Maharashtra Payment of wages Rules, 1963 are maintained.)		
Consolidated register of unclaimed Wages & Fines	C	To maintain always
Statement of Employee's and Employer's contribution	A-1	Half yearly by 15th July and 15th January
Abstract under the Act		To display always

The Payment of Gratuity Act

Returns/Registers	Form No.	When to comply
Notice of opening of the Establishment	A	Within 30 days of rules becoming applicable
Notice of change in Name, Address, Employer, Nature of Business	B	Within 30 days of any such change
Notice of closure	C	At least 60 days before the intended closure
Display of Notice specifying the name and designation of authorized officer to receive the notices		To display always
Notification of family in relation to an employee	D	As and when required
Nominations	F	Within 30 days of completion of 1 year of service
Notice of payment of gratuity	L	Within 15 days of the receipt of the application
Abstract under the Act		To display always
Compulsory Insurance		To obtain an insurance for the liability for payment towards the gratuity

The Bombay Shops & Establishments Act

Returns/Registers	Form No.	When to comply
Application for registration	A	Within 30 days from the date of commencement
Application for renewal of registration	B	15 days before the expiry of registration

(Contd.)

Returns/Registers	Form No.	When to comply
Registration Certificate	D	To display always
Notice of change in respect of change on no. of employees	E	Within 15 days after the expiry of the quarter
Notice of change in respect of any other change	E	Within 30 days after the change has taken place.
Register of recording lime washing, color washing	F	To maintain always
Register of employment	H/J	To maintain always
Register of leave	M, N	To maintain always
Name of board in Marathi		To display always
Ascertainment of age of employee	O	To get from registered medical practitioner
Inspector's Visit Book		To maintain always
List of Weekly offs and Paid Holidays		To display

The Industrial Disputes Act: This act covers the provisions related to Change in the service condition of any workman, Strike, Lock-out, Lay-off, Retrenchment

Returns/Registers	Form No.	When to comply
Notice of change in the conditions of service	XIII	Before intending to effect any change in the conditions of service applicable to any workman.

The Water (Prevention & Control of Pollution) Act

Returns/Registers	Form No.	When to comply
Application for consent to establish	XIII	Before establishing or taking any steps for establishment of industry/operation/process/or any treatment/disposal system for discharge, continuation of discharge.
Renewal of consent to establish/operate		Before ending the period of the existing consent
Return under Cess Rules showing the quantity of water consumed in the previous month	1	On or before 5th of every month

The Employees' State Insurance Act

Returns/Registers	Form No.	When to comply
Registration of factories/establishments	1	Before 15 days of applicability of the Act
Declaration form	1	To get filled up from all employees on the date of joining
Declaration form to be sent to appropriate office	3	Within 10 days from the date of filling up of form

Registration of families	1A	Within 10 days from the date of filling up of form
Change in family declaration	1B	To be filled in by employee and employer to forward it within 10 days
Return of contributions to be send in triplicate	6	Within 42 days of the termination of the contribution period to which it relates
Challans of contributions		Monthly to submit in Bank
Register of employees	7	To maintain
Accident Book	15	To maintain
Accident Report	16	Immediately, if accident is serious or within 24 hours

List of acts and commissions which amounts to misconduct

a) Willful insubordination or disobedience, whether or not in combination with another, of any lawful and reasonable order of a superior.

b) Going on illegal strike or abetting, inciting, instigating or acting in furtherance thereof;

c) Willful slowing down in performance of work, or abetment or instigation thereof;

d) Theft, fraud or dishonesty in connection with the employers' business or property or the theft or property of another workman within the premises of the establishment;

e) Taking or giving bribes or any illegal gratification;

f) Habitual absence without leave, or absence without leave for more than ten consecutive days or overstaying the sanctioned leave without sufficient grounds or proper or satisfactory explanation;

g) Late attendance on not less than four occasions within a month;

h) Habitual breach of any Standing Order or any law applicable to the establishment or ant rules made there under;

i) Collection without the permission of the Manager of any money within the premises of the establishment except as sanctioned by any law for the time being in force;

j) Engaging in trade within the premises of the establishment;

k) drunkenness, riotous, disorderly or indecent behavior on the premises of the establishment;

l) Commission of any act subversive of discipline or good behavior on the premises of the establishment;

m) Habitual neglect of work, or gross or habitual negligence;

n) Habitual breach of ant rules or instruction for the maintenance and running of any department, or the maintenance of the cleanliness of any portion of the establishment;

o) Habitual commission of any act or commission for which a fine may be imposed under the Payment of Wages Act, 1936.

p) Canvassing for union membership, or the collection of union dues within the premises of the establishment except in accordance with any law or with the permission of the Manager

q) Willful damage to work in process r to any property of the establishment;

r) holding meeting inside the premises of the establishment without the previous permission of the Manager or except in accordance with the provisions of any la for the time being in force;

s) Disclosing to any unauthorized person any information in regard to the processes of the establishment which may come into the possession of the workman in the course of his work;

t) Gambling within the premises of the establishment;

u) Smoking or spitting on the premises of the establishment where it is prohibited by the employer;

v) Failure to observe safety instructions notified by the employer or interference with any safety device or equipment installed within the establishment;

w) Distributing or exhibiting within the premises of the establishment hand-bills, pamphlets, posters, and such other things or causing to be displayed by means of signs or writing or other visible representation on any matter without previous sanction of the Manager;

x) Refusal to accept a charge-sheet, order or other communication served I accordance with these Standing Orders;

y) Unauthorized possession of any lethal weapon in the establishment.

Authorities under various Labor Laws

1.	The Factories Act.	Joint Director Industrial Safety & Health
2.	The Minimum Wages Act.	Joint Commissioner of Labour.
3.	The Payment of Wages Act.	Joint Commissioner of Labour.
4.	The payment of Bonus Act.	Joint Commissioner of Labour.
5.	The Contract labor Act.	Joint Commissioner of labor.
6.	The Pollution Control Act.	The Sub Regional Officer, MPCB.
7.	The E.S.I.Act.	The Sub Regional Officer ESIC.
8.	The Employment Exchange Act.	The Sub Regional Officer.
9.	The Bombay Labor Welfare Act.	The Labor Welfare Officer.
10.	The Shops & Establishment Act.	The Assistant Commissioner.
11.	The Workmen's Compensation Act.	Joint Director Industrial Safety & Health.
12.	The Provident Fund Act.	Commissioner Provident Fund.

COMPLIANCES

Applicability of Acts

The Factories Act

Applicability: Any premises including precincts thereof

- where ten or more workers are working or were working on any day of preceding twelve months and in any part of which a manufacturing process is being carried on with the aid of power,

- where twenty or more workers are working or were working on any day of preceding twelve months and in any part of which a manufacturing process is being carried on without the aid of power.

Scope: An Act to consolidate and amend the law regulating labor in factories

Comments: This Act covers basic rules and regulations for starting of a new factory and maintaining it consistently under the conditions prescribed in the Act for safety and health and welfare of the employees.

Summary of Factories Act

The most critical act

1. Occupier & Factory Manager are liable for most violations

2. Each unit must have valid factory license

3. Renewal application to be sent on time (60 days before expiry in most states)

4. Total number of workers not to exceed number specified in License

5. Stability certificate to be renewed every 5 years

6. Pressure Vessels to be checked every 6 months/1 year

7. Lift, Hoists & Tackles to be checked every year for load test & safety

8. Employees medical to be undertaken once every year.

Other conditions for Factories Act

- Follow all Health & Safety regulations

- Crèche required for more than 40 women workmen subject to certain rules

- Canteen requirement

- Provision for clean, safe, adequate drinking water

- Adequate provision for toilet facility

- Special regulations in case any children working

Forms Related: (Note that this is not standard implementation in all the sates in India. It may vary slightly depending upon the State as there may be additional State Factory Rules framed post Factory Act. For e.g. Form 1-A in Punjab will be called as Form 1-B and Form 1-A is used as Questionnaire for rooms in the Factory. Implementation is a state subject.)

FORM NO. 1	APPLICATION FOR PERMISSION TO CONSTRUCT, EXTEND, OR TAKE INTO USE ANY BUILDING AS A FACTORY IT WILL INCLUDE SUBMISSION AND APPROVAL OF PLANS
FORM NO. 1-A	CERTIFICATE OF STABILITY OF A FACTORY OR A PART OF A FACTORY (Once every 5 years)
FORM NO. 2	APPLICATION FOR REGISTRATION & GRANT OR AMENDMENT OF LICENSE
FORM NO. 3	GRANT AND RENEWAL OF LICENSE AND NOTICE OF OCCUPATION (Renewal is usually done at the end of Calendar Year – 31st Dec)
FORM NO. 4	LICENCE TO WORK A FACTORY (Issued by the State Government)

Returns/Registers	Form No.	When to comply
Notice of Change of Manager	5	As and when Factory Manager changes
Certificate of Fitness	6	
Health Register	7	It is a state subject (may/may not exist) and needs to be complied annually.
Record of lime washing, varnishing, painting, etc. & repainting, re-varnishing	8	To be maintained always
Humidity Register	9	
Register of workers attending to machinery	10	To be maintained always
Prescribed for Report of Examination of Hoist or Lift	11	
Prescribed for report of examination of the lifting machines, ropes and lifting tackles	12	Certificate to be obtained annually
Examination of pressure plant by competent person	13	Every Six Months
Externally		Once in Twelve months
Internally		Once in four years
Hydraulic Test		
Register of Compensatory Holidays	14	To be maintained always
Notice of period of work for adults	16	To display and maintain
Register of adult workers	17	To be maintained always
Notice of periods of work for child workers	18	
Register of Child Workers	19	

Leave with wages register	20	To be maintained always
Leave Book	21	To be maintained always
Prescribed under Schedule II, III, IV, VIII, XI, XVII, XVIII, XX Special Certificate of Fitness	23	
Report of accident by the Manager	24	To be submitted to Factory Inspection office as and when accident takes place
Notice of Dangerous Occurrence	24A	Within 12 hours of taking place of such accident
Notice of Poisoning of Disease	25	
Abstract under the Act	26	To be displayed always
Annual Return	27	To be submitted to Factory Inspection office on or before 1st Feb. every year
Muster Roll	29	To be maintained if Form 17 & 19 are not maintained
Register of accidents & dangerous occurrences	30	To be maintained always
Inspection Book	31	To be maintained always
Notification of Paid Holidays		To display and to submit to Factory Inspection Office every year in January

General Permits required for any Factory or Workshop or warehouse setup

- MSME (DIC)/LSE (SIA) Registration
    - Part – I
    - Part – II
- Environment:
    - RRZ Policy – River Distance Certificate from State Irrigation Department
    - Environmental Clearance
    - State Pollution Control Board
        a) Consent to Establish
        b) Consent to Operate
- Petroleum and Explosives Act
    - Gas Cylinder Rules
    - Storage of Diesel (DG Set)
- Town Planning Approvals
- Shop and Establishment Act/Factory Act
- Fire NOC
- Water and Electricity Permits
- MSME – DIC/LSE - SIA Registration:

- MSME – Micro/Small/Medium Enterprise Registration – EM Part A and Part B from District Industry Center
- LSE – Large Scale Enterprise Registration – IEM – Part A and Part B at Secretariat of Industrial Assistance

Investment Ceiling for Plant and Machinery

Classification	Manufacturing Sector	Services Sector
Micro Enterprise	Up to Rs 25 Lacs	Up to Rs 10 Lacs
Small Enterprise	More than Rs 25 Lacs and does not exceed Rs 5 Crores	More than Rs 10 Lacs and does not exceed Rs 2 Crores
Medium Enterprise	More than Rs 5 Crores and does not exceed Rs 10 Crores	More than Rs 2 Crores and does not exceed Rs 5 Crores
Large Enterprise	More than Rs 10 Crores	More than 5 Crores

USE OF MSME/LSE

✓ Record Keeping, Pre-Requisite for other Permits/Licenses, Checks and Balances for Industrial Licensing

✓ Benefits and Schemes for MSME's – ISO Quality Audits & Certifications, CRISIL Ratings, International Exhibition, Training etc.

Environment – RRZ Policy – River Distance

- RRZ Policy formed by Environment Department, Government of Maharashtra to tackle pollution, low rainfall, low water flow rate in Rivers, rapid industrialization and many other reasons.

- Classes of River : A-I : from origin up to the first dam (drinking water), A-II : from first dam to designate (domestic use), A-III : suitable for fisheries and wildlife, A-IV : suitable for agricultural and industrial usages

Classification of River

Classes	No Development Zone	Only Green category of Units with pollution control devices.	Only Orange category of Units with pollution control devices.	Any type of industries (Red, Orange, Green) with pollution control devices
A-I	3 Km on the either side of river	From 3 Km to 8 Km from river (H.F.L.) on either side	From 3 Km to 8 Km from river (H.F.L.) on either side	Beyond 8 Km from river (H.F.L.) on either side.
A-II	1/2 Km on the either side of river.	From 1/2 Km to 1 Km from (H.F.L.) on either side	From 1 Km to 2 Km from (H.F.L.) on either side	Beyond 2 Km from river (H.F.L.) on either side.
A-III	1/2 Km on the either side of river	From 1/2 Km to 1 Km from river (H.F.L.) on either side	From 1/2 Km to 1 Km from river (H.F.L.) on either side	Beyond 1 Km from river (H.F.L.) on either side.
A-IV	1/2 Km on the either side of river	From 1/2 to 1 Km from river (H.F.L.) on either side	From 1/2 to 1 Km from river (H.F.L.) on either side	Beyond 1 Km from river (HFL.) on either side.
MIDC with CETP (Exemption)	1/2 Km on the either side of river	From 1/2 Km to 3/4 Km from river (H.F.L.) on either side	From 1/2 Km to 3/4 Km from river (H.F.L.) on either side	Beyond 3/4 from river (H.F.L.) on either side.

✓ MPCB and State MoEF will maintain the classification of Rivers and will monitor and update the same on continuous basis.

✓ State Irrigation Department will issue the River Distance Certificate after measurement of distance of proposed unit from the nearest river/dam/reservoir.

Environment – EC and EIA

◆ Environment clearance mandatory for Real Estate Projects above 20,000 Sq. Mtrs. in India.

◆ Environmental Impact Assessment for Buildings above 20,000 Sq. Mtrs. The EIA is a comprehensive assessment of resource use, including energy, water, air, land, and ecological impacts.

◆ In 2009, the central government made it mandatory for any new construction activity in India to procure environmental clearance from the Union Ministry of Environment and Forest (MOEF) or its state-level body.

◆ The built up or covered area on all the floors put together including basement(s) and other service areas, which are proposed in the building/construction projects. As per the directive, the developers, after acquiring the environment clearance, have to move their files to urban department of the state for building plan approval.

◆ Why is Environmental Impact Assessment above 20,000 Sq. Mtrs for Buildings mandatory? In the Indian context, 20,000 Sq. Mtrs. is a fairly large area. It amounts to 200 apartments in a composite apartment building and the project population would be 1,000 which would require 140 kilo liters of water supply a day and 112 kilo liters of sewage treatment per day.

MPCB Consent to Establish and Operate

◆ MPCB Consent to Establish and MPCB Consent to Operate is mandatory if any of the following is applicable:

  - Construction of building more than 20,000 Sq. Mtrs. (built up)

  - Environmental Clearance is required (CRZ Norms/Forest Act/Environment Act/RRZ Violation)

  - Installation of DG Set – More than 1MW (Less than 1 MW, it would be Consent/NOC from MPCB under Air Act and Noise)

  - Hazardous Waste Generation

  - Orange Category Unit for IT & ITeS Activity (Size not notified by MPCB)

  - Orange Category Unit for testing and laboratories. Can be classified as Red based on pollution if recommended by the Board

  - Hospitals/Clinics - Bio Hazardous Waste Generation

State Pollution Control Board Compliances

◆ Ensure renewal of PCB Consent to Operate on regular basis. (i.e. 1 year/2 years/3 years etc. depending upon the category of the Consent)

◆ Understand the entire PCB Consent and ensure compliance. (e.g. 33% open area under plantation etc.)

Understand the following

✓ Non Hazardous Solid Waste (Municipal Waste, Recycler Waste etc.)

✓ Hazardous Waste (Oil Soaked Cotton Waste, Oil Drums etc.) - Dispose to MEPL/MWML

✓ Bio Hazardous Waste (Discarded Needles, Blood Samples etc.)

- Ensure proper accounting of Waste (Municipal Solid Waste – Internal Record Keeping, Hazardous Waste – Form 13 Manifests)

- Ensure STP/ETP Tests from MoEF Approved Laboratory every 3 to 6 months.

- Ensure DG Set Stack tests and spot noise tests from MoEF Approved Laboratory every 3 to 6 months.

- Ambient Air Monitoring Tests (National Ambient Air Quality Standards), Work zone Air – as per factories act standards.

- Ensure WATER CESS returns to MPCB on monthly/by-monthly basis as per process defined by MPCB and when the water bills are generated by Municipality/Agency.

- Ensure WATER CESS payments to MPCB – Monthly/by-Monthly/half yearly/annually – depends upon the amount.

- Ensure Annual Returns – Form V Environment Statement and Form 4 Hazardous Statement before the deadlines.

HOUSEKEEPING

HK Basics: Cabin/ Office Cleaning

➤ **Dry Dusting**

Introduction: Dry Dusting is the action/skill of removing loose dirt from any surface.

Advantages: Regular/periodic dry dusting gives commendable results.

Disadvantages: Improper dusting displaces dirt (does not remove) and sometimes results in scratches on the surface being minute granular in nature.

Objective/Purpose: To Clean and maintain applicable surface in an effective manner.

Equipment Required: Dry Dusters (2 No's).

- The area is to be divided into smaller sections if the area for dusting is large

- Based on the activity that is to be carried out the appropriate duster is to be utilized. (soft/medium soft/rough texture)

- Dry Dusting is often followed by damp dusting to remove the left behind fine particles of dust.

- It is recommended to start with the furthest end of the area allocated and complete toward the entrance either working in clockwise or anticlockwise direction based on the area layout so that no area is left unattended.

- Dusting is to be done from corner to corner ensuring all areas are covered. Removable items are to be lifted/ removed for dusting.

- Each dusting cloth if used correctly & wisely can be utilized for a longer period. It is generally recommended to use the eight sides of a duster to good use, ensuring hygiene is not compromised.

- The latest efficient duster though a bit expensive is preferably the Micro fiber cloth.

➤ **Damp Dusting**

Introduction: Damp Dusting is the action/skill of removing dirt/soil left behind after the Dry Dusting activity has been completed from any surface.

Advantages: Regular/periodic Damp Dusting gives commendable results, the life of the product/surface is increased and the area is well maintained.

Disadvantages: Improper usage of Damp dusting process could also damage certain delicate/sensitive surfaces. (E.g.: wood Surfaces).

Objective/Purpose: To Clean and maintain applicable surface in an effective manner.

Equipment Required: Dusters (2 No's), Portable small Blue and Red buckets (Stationed in the trolley).

- The area is to be divided into smaller sections, if the area for dusting is large.

- Based on the activity that is to be carried out the appropriate duster is to be utilized. (soft/medium soft/rough texture)

- Damp Dusting is often followed after dry dusting to remove the left behind fine particles of dust.

- It is recommended to start with the furthest end of the area allocated and complete toward the entrance either working in clockwise or anticlockwise direction based on the area layout.

- The Dusters are rinsed in the Red Bucket containing plain water first to remove any soil and thereafter rinsed in the Blue bucket containing diluted detergent (As required) and the excess solution is squeezed into the Red bucket.

- Damp Dusting is to be done from corner to corner ensuring all areas are covered. Removable items are to be lifted/removed for dusting.

- Each dusting cloth if used correctly & wisely can be utilized for a longer period. It is generally recommended to use the eight sides of a duster to good use, ensuring hygiene is not compromised.

- Diluted detergents or plain water is used to act as a magnet to attract/make the left behind loose particles of dirt to stick on to the dusting cloth.

- At times a little agitation is required to suspend the soiled particles for removing it efficiently.

- The latest efficient duster though a bit expensive is preferably the Micro fiber cloth (Detergents are not to be used which saves cost).

➤ **Dry Mopping**

Introduction: Dry Mopping is the action/skill of removing dirt/soil from any floor surface using a dry mop.

Advantages: Regular/periodic Dry Mopping gives commendable results, the life of the floor/surface is increased and the area is well maintained

Disadvantages: Improper procedure of Dry Mopping displaces dirt (does not remove) and results in scratches on the surface, being minute granular in nature (Especially Marble and wood)

Objective/Purpose: To Clean and maintain floors in an effective manner

Equipment Required: Dry Mop (Ezee mop/Poly acrylic Mop), Vacuum Cleaner and Cleaning in Progress sign board.

- The area is to be divided into smaller sections if the area for Dry Mopping is large. It is recommended for a caution Sign board to be placed before commencing work.

- Based on the activity/floor the appropriate dry mop is to be utilized. (Acrylic,   normal cotton piles, lamello, Microfiber refills)

- Dry Mopping is often followed by damp mopping to remove the left behind fine particles of dust.

- It is recommended to start with the furthest end of the area allocated and   complete toward the entrance. A zigzag or straight mopping flow is carried out based on the area layout, traffic foot falls.

- Dry Mopping is to be done from corner to corner ensuring all areas are covered. Removable items are to be lifted/removed for the process.

- Each dry mop if used correctly & wisely can be utilized for a longer period. It is generally recommended to use the mop just before the soiled particles falls off without adhering to the surface, which requires experience to judge, ensuring standards are not compromised.

- It is recommended to vacuum the Dry Mop head to ensure 90 % of soiled dust particles are disposed efficiently rather than the Dustpan/Brush method.

- Acrylic and Micro fiber Mops are used to act as magnets to attract /make the left behind loose particles of dirt to stick on to the Dry Mop.

Note: It is recommended to have adequate stock of dry Mop refills based on frequency of change, area, traffic foot falls.

➢ Wet Mopping:

Introduction: Wet Mopping is the action/skill of removing adhered dirt/soil from any floor surface using a wet mopping process.

Advantages: Regular/periodic Wet Mopping gives commendable results, the life of the floor/surface is increased and the area is well maintained.

Disadvantages: Improper procedure of Wet Mopping results in damaging the floor surface, (Especially Marble and wood).

Objective/Purpose: To Clean and maintain floors in an effective manner

Equipment Required: Wet mop, (Color coded) double bucket trolley filled to 30 % of its capacity with fresh water in the red bucket and 70 % of its capacity with diluted detergent as per labeled specifications in blue bucket respectively and Cleaning in Progress sign board.

- If the Wet Mopping area is large it is to be divided into smaller sections.

- A caution Sign board is placed before commencing work.

- Based on the activity/floor the appropriate Wet mop is to be utilized. (Kentucky Mop, lamello refills, etc…).

- It is recommended for mops refills to be color coded for hygienic standards.

- Wet Mopping is often followed after Dry mopping to remove the left behind adhered fine particles of dust, dirt, it also disinfect and improves the cleanliness and hygiene standards based on the detergent used.

- A double bucket (Color Coded) mop trolley is used, the Red bucket is filled to 30 to 40% of its capacity. The blue bucket is filled with diluted detergent to about 70 to 80% of its capacity.

- The clean mop is first rinsed in the Red bucket to remove any soil and squeezed in the wringer above, later it is again rinsed in the blue bucket and the excess liquid is squeezed into the Red bucket. The floor is mopped with the required amount of dampness as required.

- It is recommended to start with the furthest end of the area allocated and   complete toward the entrance. The sides of the area are completed as if boxing in an rectangular area for Mopping.

- The mopping flow is to be that of an overlapping "8" completing the area corner to corner.

- Each wet mop if used correctly & wisely can be utilized for a longer period. It is generally recommended to use the mop until its color changes from that of white to light Ash, after which it is required to be washed. (requires experience to judge, ensuring standards are not compromised)

- Color coded mops are washed separately to ensure standards of hygiene are   maintained.

- Note: It is recommended to have adequate stock of Mop refills based on frequency of change, area, traffic and foot falls.

➢ Glass Cleaning Caddy Kit

   ✓ Glass Cleaning Caddy

   ✓ Mop Cloth – 2 no's

   ✓ Yellow wipe Sponge

   ✓ Vise-Versa window Squeeze/ applicator

   ✓ Glass Cleaning spray Can

             (or)

   ✓ Glass cleaning Bucket (As required)

   ✓ Glass Cloth (Buffing)

➢ Window and Glass cleaning

- Fill the glass cleaning bucket with diluted detergent as per labeled Specs.

- Proceed to the allocated work area with the required material/tools as detailed in the kit description. Divide the area into sections if the area is large.

- Place 2 to 3 Mop cloths as required preventing dripping of detergent on the floor/carpet.

- Start cleaning from the top to bottom. First clean the glass starting from the Frame bottom.

- Apply diluted detergent with applicator and utilize the vise-versa window squeeze for cleaning.

- Wipe the edge of the frame and glass corners with the yellow sponge wipe cloth.

- Use telescopic pole as required to reach areas not accessible otherwise.

- After completing cleaning of each section of the glass, wipe/mop the dripped residue from the floor using a floor cloth.

- Buff with a glass cloth to get the required shine avoiding streaks.

Cabin/Office/Conference Cleaning

Objective/Purpose: To maintain a clean and pleasant work environment.

Machine/Equipment Required:

Janitorial trolley with red, blue buckets, garbage bags, bin liners, trigger spray cans, all purpose cleaner, glass cleaner, air freshener, blue sponge wipe cloth, Yellow cleaning cloth, sponge scrubber, drying cloth, Keyboard cleaning brush, tissue box, Vacuum machine, (If required) and Cleaning in Progress sign board.

Procedure:

- Proceed to the work area with the required items and materials.

- Place the Cleaning in Progress sign board at the work area, switch on lights and check for non-working appliances which are to be reported to the helpdesk or the maintenance control desk for rectification.

- Locate the trolley and vacuum cleaner near the entrance of the office and ventilate office if applicable at all times.

- Empty the bins and ashtrays into the garbage bag fixed on the trolley.(Make sure the cigarette butts are extinguished and change the bin liners if they are smeary).

- Start from one side of the office either following the clockwise or anticlockwise method to ensure that no areas are left out and till you reach where the work had begun.

- Take the blue sponge wipe cloth, soak in blue bucket (diluted detergent as per labeled instructions), wring well and start wiping table top from one side working along the table by moving objects, files one at a time and placing them back in place. Continue the process until the entire table is cleaned.

- Clean complete monitor screen with a lint free yellow flannel cloth and the keyboard with a brush and damp wipe the monitor. Ensure wires are not disconnected/damaged or altered during the cleaning process.

- Wipe the telephone with the wet sponge cloth, with disinfectant/deodorizer. Wipe the bases of tables and chairs also with the damp cloth.

- Wipe the top and vertical surfaces of all partitions, cabinets and shelves in the office with a damp blue sponge.

- Enamel painted walls can be cleaned using the method applicable for glass cleaning

- After the cleaning activity is completed, spray air freshener as required and close the door.

- Clean all items/materials used and ensure that the vacuum machine's dust bag is replaced if necessary.

- As per requirement of the area, the activity is predetermined and scheduled to ensure that the standards are maintained.

Do's (Basic HK)

Do's

- Report to work 5 to10 minutes before the shift, wearing your complete uniform.

- Attend the briefing and listen carefully to the instructions that are being passed.

- Ask relevant questions if you should need any clarification.

- Take a round of your area and report snags/problems if any.

- Ensure you are stocked with the required tools and equipment's to carry out your work, if not report it to your Supervisor.

- Honesty, sincerity, dedication and integrity at work are traits valued by Knight Frank.

- Every employee shall communicate with the client in a courteous and pleasant manner. Wish/greet the client in a soft and audible manner.

- Employees shall adhere strictly to the Code of Conduct as laid down by the client and By KF.

- Employees shall also follow the laws enforced by the regulatory body/Legal Bodies of India.

- You are to safeguard the property of the client as your own. If some valuable items are left unattended, inform the same to your superior or helpdesk. Lost and found items are to be safely handed over to the client or reported accordingly to your superior.

- If you have any doubts or clarifications, address them to your superior accordingly.

- Inform you're superior if you complete your work before the stipulated time to enable him to detail you to carry out additional task for which you would be recognized and appreciated.

- Complete the checklist as required and report to your superior before you leave your work area.

- Intimate your superior well in advance if you require leave, Should you also have any problems reporting for work, intimate your superior at the earliest so that timely alternative arrangements can be made.

- Sport a confident and smiling face; enjoy your work from the depth of your heart.

Don'ts (Basic HK)

- Do not misuse tools, equipment and company property.

- Do not accept or solicit any kind of gifts, donation or bribe either from the client or from the vendors.

- Do not solicit or explore employment opportunities with the client while working for Knight Frank or the Vendor.

- Do not shout or raise your voice in the work premises. Talk softly and perform your work quietly in an un-obstructive manner.

- Do not disclose any information about the client or the company to outsiders. (Non-Disclosure agreement)

- Do not leave your work/area without being relieved. Should you have to leave for some reason, inform your superior.

- Do not bad mouth/gossip/or form groups in work premises.

- Do not steal or pilfer any items/materials of the client or premises.

- Do not remove anything from the site without the prior approval from the client.

- Do not use drinking water Bottles, other cups & containers for the use of Detergents other than the specified Labeled containers

- Do not visit websites which display content that is indecent/obscene or explore employment opportunities.

Floor Care

Carpet Flooring

As lovely as carpeted flooring looks and feels, it is not quite easy to maintain. You will need to a lot of care if you want to keep it as good as new.

Dust tends to accumulate in the fur of the carpet so you must vacuum it every alternate day. And here's a handy hint- make sure you do it in the opposite direction of the pile.

If your carpet happens to get stained, a toothbrush and mild soap (even your liquid hand soap) should do the trick. You can even shampoo your carpet with a mild detergent or hair shampoo if need be.

In case one spills a little water, just blow-dry the carpet or leave it to dry on its own.

Tip

- Opt for a removable carpet. This way you can have it washed once a year. And don't you worry, a carpet cleaner will do it for you.

Wooden Flooring

If you want to give your room that cozy feeling, there is no better way than with wooden flooring. But when it comes maintaining those floors, here is a tip or two to help you.

Don't drag heavy furniture around the room. It will cause scratching and denting in the wood. Tiny felt pads placed under the legs of your furniture are a great way to ensure the floor doesn't get ruined.

You might also want to watch your step with those stilettos. If there are little stone particles on the heel, your precious wooden flooring could get damaged.

A regular damp mopping should keep the flooring clean and in good condition. But avoid excess water.

If anything liquid spills, make sure you clean it up quickly. You don't want it to seep in and destroy the wood.

Routine Maintenance

It's a delegate task to maintain the wooden flooring. For a routine cleaning maintenance one is required to use a soft bristle broom, and vacuuming with a soft floor attachment if your wooden floor has any beveled edge that could collect waste. As a Facility Manager you should ensure that a professional wood floor cleaning product recommended by a supplier or manufacture is periodically used.

You can also do following things to maintain the beauty of your wood floors as well.

- Do not use products that are not recommended by the manufacturer for cleaning or maintenance of wooden flooring. You should also avoid usage of wax for floor polishing.

- Never use mop & water on the wooden floors.

- In case of any kind of spillage ensure that the same is wiped up immediately with a slightly dampened towel.

- Ensure that a fabric faced glide or soft plastic under the furniture legs are placed to avoid any kind of scratching on the floor.

- Use foot mats at both side of the entrance i.e. inside and outside doorways to avoid entry of dirt, grit & other debris from being carried onto the wooden floors. This will help in preventing scratches.

- Never ever apply access wax on a wooden floor. If you find a waxed floor is getting dull, then you need to do buffing for the floor. You also need to ensure that there is no wax buildup under furniture and other light traffic areas by applying wax in these spots.

- People should avoid walking on your wood floors with cleats, sports shoes and high heels.

- While shifting or relocation of heavy furniture, one should ensure that they are not dragged on the wooden flooring.

- Use a humidifier throughout the winter months to minimize gaps or cracks.

Best Practice for Maintenance of Wooden Flooring

- Ensure that the wooden flooring is applied with lacquer or melamine by the professional agency.

- After every five years consider Re-sanding and re-polishing for your wooden floor every five.

Laminate & Vinyl Flooring

The laminate or vinyl flooring doesn't require any maintenance. A regular damp mopping would do just fine for laminate or vinyl flooring.

Warning: Never use harsh detergents or rough abrasive cleaning tools like steel wool on the flooring as it might result in irreparable damage.

Best Practice for Maintenance for Laminate & Vinyl flooring

In case by any chance the flooring gets stained badly with tar or paint, you can use acetone or nail polish remover to get rid of the strains and then swab the floor with a damp mop.

Rubber Flooring

The advantage of installing flooring is that it adds a safety element to almost any situation that they are used in and also the person who walks on the mat doesn't slip.

If you want that the rubber flooring lasts long ensure that the floor is regularly vacuum cleaned to remove all the dust and dirt and post vacuuming, you can use a damp mop for mopping.

Precaution: One has to ensure that no sharp objects fall on the floor as it could cut into the flooring, ruining it for good.

If the rubber flooring gets damage and have developed chip effect, in that case you have to be very careful about sand particles getting into the grooves. So take the added precaution of placing a floor mat outside the room for guests to wipe their feet on.

Tiled Flooring

The best, economically and easily maintainable flooring is the tiles. Tiles are of various types, they are ceramic or vitrified.

While installation of tiles one has to ensure that the grouting between tiles is done by an expert labor. This will help in reducing gaps and the chances of dirt getting stuck in them. The housekeeping team to ensure a non-acidic cleaner is used to prevent the grouting from coming off.

In case of glazed ceramic tiles, one should avoid vacuum cleaning. Sweeping is a better option as vacuum cleaning end up scratching the tiles.

Tip

Wiping of tiles with a solution of white vinegar and water in a ratio of 1:4 once in a week will keep them glowing.

Marble Flooring

The marble flooring gives better looks. You will have to take utmost care in maintaining the marble. However once scratched, you may have to replace the complete piece of marble. To avoid damage you should ensure:

- Avoid dragging furniture or other heavy objects across the floor as it will scratch or chip the stone slabs.
- The stone flooring is porous hence if by chance anything falls on the floor ensure to mop it quickly.
- The floors to be dusted and damp mopped on a daily basis. You can also use a vacuum cleaner. Just be careful not to scratch the floor.

Tip

Never use cleaning products that contain any acid content like lemon juice, vinegar or ammonia based cleaners. Acid takes away the luster of the stone. If the concentration of acid is very high, the stone could even get corroded.

Marble Floor Maintenance Guide

Marble floor needs proper attention for long lasting maintenance. If given proper attention on routine maintenance over the years, marble floors will provide historic continuity and lasting service.

Daily Maintenance

Marble is a very soft stone and is highly porous. It is more porous than granite. Because of this, it gets easily damaged by acid or any material which has acid content such as tomato & Orange juice, vinegar etc. Marbles are also prone to develop water stains because of mineral content of water or water spots after mopping.

The best way of maintaining a marble floor is daily housekeeping. One should ensure a door mat is kept at the entrance which shall absorb the dust. You should ensure to keep the floor clean of superficial dirt by using a minimum amount of plain water (warm is best) and a cotton string mop. The mopping of floor should be carried out at frequent intervals; this will prevent soil from penetrating the surface. If there is any kind of spillage one should ensure that the same is blotted out quickly.

Note: Never use detergent based cleaning chemicals, it may damage the floor.

In case there are some strain marks which needs to be removed you can make use of following measure:

a) Add 0.005% of ammonia to water (ensure it's so less amount that no odor is present) when diluted use it for cleaning.

b) Sometime one should use fluffy towel, and use it to dry the floor thoroughly instead of allowing it to dry it off naturally.

c) Once in a quarter you should get the marble floor polished by a contractor.

It's important to dry up after you clean marble tiles, as they tend to spot and stain if they remain wet.

In case the floor gets severe scratches or badly discolored; it can be restored by wet sanding and chemical stripping. Sanding needs to be followed by honing and polishing. However repeated heavy sanding can cause wear down of a floor and reduce its life. Hence it is better to avoid sanding procedure by maintaining the polished finish.

Marble floor lamination/sealer

The floor sealers or laminate is applied on marble flooring to protect it from inhibit penetration of dirt, food, and beverage stains. This also makes the marble attractive and slips resistance.

Disadvantage of sealer: It gets easily stained from water and salts during wet weather. It also tends to darken white marble.

It is advisable to leave marble floors in their natural state without coatings.

Cleaning Chemicals

There are four major types of cleaning agents commonly used by housekeepers in housing or commercial properties. Each of the cleaning agents has a specific purpose and should only be used as intended, in case of misuse it may cause dangerous and costly impacts.

The four types of cleaning agents used in housekeeping are:

1. Detergents

2. Degreasers

3. Abrasives

4. Harekrishna!192797Acids

Detergents

Detergents are substances containing soaps and/or surfactants (any organic substance/mixture) that are used for washing or cleaning jobs for the household, institutional or industrial purposes, including:

- Dishwashing

- Handwashing

- Laundry washing

- Fabric softeners

- All-purpose cleaners

- Bleaching

There are many cleaning products containing detergents and they come in various forms, including powders, tablets, concentrated liquids, liquid capsules, pastes or cakes.

Degreasers

Degreasers are used for heavy-duty cleaning to remove grease, grime, dirt and oil from hard surfaces. They are used in commercial kitchens. To remove grease from grills, ovens, and other metal surfaces. As well as from heavily soiled floors.

Working Principal of Degreasers

Grease, oils and fats are organic dirt which is broken down by alkaline solutions and solvents. Heavy-duty degreasers such as oven cleaners have a high pH (more alkaline). All-purpose cleaners for light dirt and dust have neutral pH.

Comparison of cleaning & degreasing agents on the pH scale

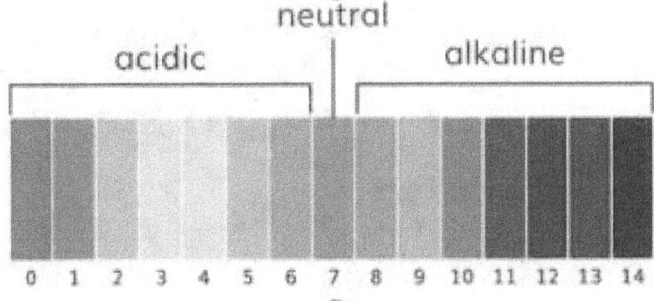

- All-purpose cleaners   pH 6-8 (neutral)
- Cleaner-degreasers   pH 9-10 (alkaline)

- Heavy-duty degreasers   pH 11-13 (high alkaline)
- Oven cleaners   pH 14 (extremely alkaline)
- Ingredients in Degreasers
- Sodium Carbonate (soap ash)
- Sodium Meta Silicate
- Ethylenediaminetetraacetate (EDTA)
- Sodium Tripolyphosphate
- Kerosene
- Methylated spirits/white spirit
- Xylene

Safety precautions while using degreasers

Before using any cleaning chemicals please read the instructions for use mentioned on packaging and follow the same strictly to avoid any ill effect on users.

High-alkaline cleaners and degreasers can cause chemical burns to the skin, hence it is important to ensure protective rubber gloves are used for handling the degreaser and it is also to ensure that the degreasers should be used in well-ventilated areas.

Some degreasing products contain ammonia or lye, which should never be mixed with bleach. As ammonia & bleach reacts with each other and can produce poisonous chlorine gas.

Alternate Safe Chemical for Degreasing

A variety of naturally alkaline ingredients can be used in degreasers in place of strong chemicals to reduce health risks. Environmentally-friendly, non-toxic and non-fuming degreasers are becoming more popular in commercial kitchens. To prevent chemical contamination.

Abrasive kind of cleaning chemicals

Abrasives are either powders or liquids used to wear off dirt from hard surfaces. These chemicals contain small particles or minerals. The effectiveness of an abrasive agent depends on the coarseness of those particles. This can be things such as sinks, floors, kitchen, and bathroom surfaces.

Below are some of the other substances used in abrasive cleaners.

- Aluminum oxide
- Calcium carbonate
- Calcite
- Feldspar
- Quartz
- Silica
- Whiting (powdered chalk)

Precautions to be adhered while using abrasive chemicals

Housekeepers should always read the labels on abrasive cleaning agents. To make sure they use the appropriate product for the cleaning task at hand. Coarse abrasives can damage surfaces such as plastic, fiberglass, glass, non-stick cookware. As well as painted woodwork, plated metal, and highly polished metal.

Housekeepers should also exercise caution when repeatedly using abrasive cleaners on hard surfaces such as sinks, bathtubs, and kitchen appliances. Because the abrasive agent gradually scratches the finish of these items.

The scratches become deeper over time and dirt becomes more deeply embedded. Requiring even stronger abrasives to clean out embedded dirtier and stains over time.

Abrasives cum Disinfectants

There are some abrasives chemicals available in market which also contain disinfectants. The purpose behind this combination is to ensure cleaning along with killing of bacteria's. These abrasive disinfecting (antimicrobial) agents can include the following:

A) Pine oil,

B) Quaternary ammonium compounds, or sodium hypochlorite (household bleach).

As chemical antimicrobial agents are regulated, the product will be labelled "disinfectant". Hence it is important for an housekeeper to closely follow the directions on the label.

Precautionary testing of Abrasive Chemicals before use

1. Always test abrasive cleaners on a small, inconspicuous area of the surface to be cleaned before using the cleaner on the entire surface.

2. Use sparingly and rub gently.

3. Do not use abrasives on marble or other natural stone surfaces.

4. Only use on surfaces not harmed by mild abrasives or acids

5. Don't allow abrasives to dry out on the surface being cleaned

Acid Cleaning Agents

Acid cleaning agents are highly corrosive and dangerous in nature. A professional approach is needed while handling and using these chemicals. These cleaning agents are often highly concentrated solutions that are used for the toughest cleaning jobs to dissolve mineral deposits (descaling).

As these chemicals are dangerous in nature hence they should be diluted according to the manufacturer's instructions before use.

Area where Acid Cleaning Agents can be used

Acid cleaning agents are normally used for following purpose:

- Descaling mineral deposits
- Rust removal

- Tough stain removal

- Dissolving

- Cleaning masonry

- Mould removal

- Bathroom tile cleaner

- Restoring tarnished or discolored metal

Every cleaning jobs which needs acid as cleaning agent require different strengths or dilutions of acid solutions. Acid cleaners are often used by housekeeping staff for cleaning bathrooms and dishwashers.

Organic Mild Acid Cleaner

The mixture of vinegar and lemon juice are mildly acidic (about 5%) and have the benefit of being organic. They can be used to remove hard-water deposits from glassware.

Strong Acid Cleaners

The acids which forms part of strong acid are Sulfuric acid, Oxalic acid & Hydrochloric acid. They can be used as rust removers and toilet bowl cleaners. These chemicals are highly toxic and poisonous.

Safety precautions while using Acid based cleaning chemicals

Due to toxic nature of these chemicals a housekeeper should follow label instructions while using acid-based cleaning chemicals. Acid cleaning agents are highly toxic so housekeeping supervisor must first understand MSDS of chemicals, does & don't, action to be taken in case of accidental ingestion or injection and then only should allow the housekeeper to use these chemicals.

- Do not mix acid cleaning agents with other cleaners.

- Avoid contact with skin or your eyes

- Avoid splashing or spilling Acid cleaners on other materials

- Ventilate rooms when using acid cleaners

- Ensure correct PPE such as gloves, eye glasses and mask is used while handling these chemicals.

Cleaning chemical used for other cleaning purpose

There are many challenges apart from the daily routine cleaning. These challenges can be unblocking a drain in the bathroom clogged with hair, soap, and toothpaste or a kitchen drain clogged with fat and grease, requiring a different kind of cleaning agent.

Tips for specialized cleaning

Description of Cleaning Job	Specialized Cleaning Chemical
Kitchen drain clogged with fat & grease	Sodium Hydroxide
Bathroom drain clogged with hair and soap	Sodium hypochlorite and sodium hydroxide
Showerhead clogged with mineral deposits from hard water (limescale and rust)	Citric, oxalic, sulfamic or hydroxyacetic acid to dissolve the minerals
Fabric stained with fungi, mould and mildew	Diluted liquid household bleach (sodium hypochlorite)
Glass stained with body oils	Solvents and alkaline cleaning agents
Glass stained with mineral salts	Acetic acid (vinegar)
Tarnished metal	Kaopolite (clay) or fine hydrous silica

Different Type of Cleaning Chemicals

When it comes to Cleaning chemicals/agents Taski or Diversey products are considered as the benchmark in the hospitality Industry.

There are specific products which need to be used for each cleaning requirement and these cleaning agents are given specific codes eg: R1, R2, R3 (The letter 'R' Stands for 'Room Care'.) Etc. for ease of identification, recognition and use.

TASKI CLEANING AGENTS LIST - R1 to R9	
TASKI R1/Diversey R1	Cleaning and Sanitising of Bathroom/Toilet surfaces
TASKI R2/Diversey R2	All-purpose cleaning agent/Hygienic Hard Surface Cleaner
TASKI R3/Diversey R3	For Cleaning Glass and Mirror Cleaner
TASKI R4/Diversey R4	For Furniture Polish and Cleaning/Furniture Maintainer
TASKI R5/Diversey R5	Air Freshener/Room Freshener/Bathroom Freshener
TASKI R6/Diversey R6	Heavy-duty toilet bowl/urinal cleaner for the removal of limescale, stains and other residues.
TASKI R7/Diversey R7	For removal of oil and grease from floor/Non-abrasive cream cleaner for water-resistant hard surfaces
TASKI R8/Diversey R8	Kettle Descaler - Highly effective acid based descaler for kettles, kitchen equipment, shower heads etc.
TASKI R9/Diversey R9	Fully formulated cleaner for cleaning all fittings and walls in the bathroom, sink, tub, tiles and fittings.

Description

TASKI R1/Diversey R1 is used for Cleaning and Sanitising of Bathroom/Toilet surfaces

Area to be cleaned: All bathroom surfaces, sink, tub, tiles, floors and fittings

How to Dilute:

For cleaning: 20 ml in 1 ltr. water

For sanitizing: 50 ml in 1 ltr. water

Usage of this Cleaning Agent:

- Spray directly on the surface to be cleaned
- Leave for 2 seconds
- Scrub if necessary and wipe the surface with a clean and dry cloth
- Replace cloth regularly

TASKI R2/Diversey R2: All-purpose cleaning agent/Hygienic Hard Surface Cleaner

Area to be cleaned: All types of floor and walls

How to Dilute:

Normal soiling: 20 - 40 ml in 1 ltr. Water

Heavy soiling: 50 ml in 1 ltr. water

Usage of R2:

- Floor cleaner for glass and floor like Italian marble Can be used for wet mopping as well as scrubbing with a machine.
- Wet moping solution to be taken in bucket/mop trolley
- Rinse the mop frequently
- Alternatively, use scrubbing machine and pickup direct solution using a wet vacuum cleaner

TASKI R3/Diversey R3: Glass Cleaner and Mirror Cleaner

Area to be cleaned: Windows, mirrors, glass display cases

How to Dilute:

20 – 50 ml in 1 ltr. water for cleaning all types of glasses and mirrors

Usage of this Cleaning Agent:

- Spray directly on a dry clean cloth
- Apply to the surface and wipe with a clean dry lint free cloth
- Replace cloth regularly
- Buffing dry

TASKI R4/Diversey R4: Furniture Polish/Furniture Cleaning/Furniture Maintainer

- Area to be cleaned: All wooden floors and furnishings

How to Dilute:

Ready to use, No need to dilute.

Usage of this Cleaning Agent:

- Shake the bottle well before use.

- Spray on a soft dry cloth.

- Apply to the surface evenly and start buffing.

- Buffed the floor/surface to a high shine.

- Replace cloth regularly.

Note: Do not use on glasses, floors, stairs and laminated sheets

TASKI R5/Diversey R5: Air Freshener/Room Freshener/Bathroom Freshener

Area to be used: Offices. Corridors, washrooms

How to Dilute:

Dilution not needed

Usage of this Cleaning Agent:

- Do not spray directly on the floor

- Spray upward into the centre of the room as required

TASKI R6/Diversey R6: Toilet bowl cleaner / Heavy-duty toilet bowl/Urinal cleaner for the removal of limescale, stains and other residues.

Area to be cleaned: Toilet bowls and urinals

How to Dilute:

Dilution not needed

Usage of this Cleaning Agent:

- Heavy duty toilet bowl and urinal cleaner.

- Flush around bowl especially around rim and bowl waterline.

- Direct nozzle under toilet rim and evenly over the surfaces.

- Leave for 5 – 10 min.

- Flush toilet.

- Push water level down with toilet brush.

Note: Do not use on stainless steel, enamel, marble and tiles.

TASKI R7/Diversey R7: For removal of oil and grease

Area to be cleaned: For removal of oil and grease from floor/Non-abrasive cream cleaner for water-resistant hard surfaces.

How to Dilute:

Normal soiling: 20 - 40 ml in 1 ltr. water

Heavy soiling: 50 ml in 1 ltr. water

Usage of this Cleaning Agent:

- For wet mopping, take the solution, bucket and mop.
- Rinse the mop frequently.
- Alternatively, use a scrubbing machine and pickup solution with a wet vacuum.

TASKI R8/Diversey R8: Kettle De-scaler

Area to be cleaned: Kettle De-scaler - Highly effective acid based de-scaler for kettles, kitchen equipment, shower heads etc.

How to Dilute:

Needed no dilution

How to Use:

- Pour the required amount of product into the kettle.
- Add cold water to the maximum line and leave overnight.
- Rinse thoroughly with fresh cold water.
- Boil water once and then pour away before using again.

Usage:

- Citric acid based - safe on surfaces, including plastic kettles.
- Suitable for descaling all kettles.
- Can be used neat for tough limescale deposits on shower heads/ tap bases.

TASKI R9/Diversey R9: Removal of hard stains from Bathroom Walls and Fittings

Area to be cleaned: Fully formulated cleaner for cleaning all fittings and walls in the bathroom, sink, tub, tiles and fittings.

How to Dilute:

50 - 100ml in 1 ltr. water, as per the staining requirement

Usage of this Cleaning Agent:

- Prevent scale dirt on wall fittings
- Spray directly on the surface to be cleaned
- Leave for 20 sec

- Scrub and drain plain water

- Wipe surface and polish all metal surfaces with a clean cloth

- Replace cloth regularly

CAFETERIA MANAGEMENT

Importance of Cafeteria management

- Cafeteria management has always been a critical facet of facility management while considering it as one of the most 'integral part' at the same time.

- 'Good food' with a better 'Dining facility' is one winning combination & a key to success towards employee satisfaction for any organization.

- A good cafeteria manager with the right work knowledge will always be an asset.

Cafeteria Module

- Cooking/ In-house Cafeteria – Requires a dedicated place for dining with cooking facility that should include modern kitchen equipment's & machineries, this model works to the advantage of the organization as the food is cooked in-house with better quality checks.

- Serving/ Vendor driven Cafeteria – This model is mostly preferred by the MNC's who do business on a rent & lease basis. Here too it requires a dedicated place for dining but with limited kitchenette infrastructure, mostly the food is prepared at the vendor's central/ mother kitchen & transported to the site.

Legal & Operational Requirement

- As per the Law, any entity involved in mass manufacturing of food needs to get either Registered or Licensed by FSSAI (food safety & standards authority of India) or FDA. The Caterers supplying to the Corporate Canteens and the Corporate Canteens that are in cooking their own food are no exception to the Law.

- Registration/Licensing of all Food Vendors

- Registration/Licensing of all the in-house Corporate Canteens

- Formulation of proper Food Safety Management System (FSMS) Plans - This plan ensures quality monitoring of the food produced across the entire Food Manufacturing Process. This plan needs to be reviewed annually and its implementation needs to be monitored.

- Adherence to the FSMS Plans and assurance of Food Safety

- Contract agreement/ LOI with defined SLA's on renewable basis with the vendor, it also consists specific required services details e.g. standard menu compositions, per employee/ person portioning (approx. weight approximately varies between – 120/ 150grms), rate agreement & other legal compliances – like submission of required/ renewed licenses (e.g. FDA, labor contract, minimum wages etc..) on time, PF & ESIC challans, general & employee health & liability insurances etc..

Audits, Compliance & Checklists

- The person responsible for Food Safety should be trained on evaluation of a canteen based on the checklist. He should have enough knowledge for identification of wrong practices and adhere to the Corporate Standards while filling the checklist. Help of an external professional should be sought if required

- Daily Audits needs to be conducted by the responsible employee to ensure that the necessary protective and preventive measures required for Food Safety are being followed. The checklists should be filed and documented for maintenance of records of Food Safety Management.

- On a monthly basis the Food Safety In-charge from the company should review the Daily Audit reports of the Canteens and the Vendor's Kitchen. The regular areas of non-compliance should be identified and corrected. Seniors from the company and the Vendors should also be involved in this process.

- Apart from the Internal Audits, periodic audits by Third Party Auditors must be done. These audits will enable the company to ascertain if any aspects of Food Safety are being missed untouched, and to validate the findings of the Internal Audits. The External Audit reports too should be documented.

- As validation of all the Food Safety Initiatives undertaken by the company, test reports of samples of Food, Water and Hand Swabs of the Food Handlers are to be obtained. These will validate whether the Food Safety Initiatives are effective enough.

- Policy to be followed by Canteens across all the branches/centers of the company across India. All the Food Vendors supplying food to any of the Company's canteens need to comply with the standards.

- Development of Food Safety norms and dedicated FSMS plans for handling food at each of the outsourced canteens. The plans need to be site specific to ensure relevance and proper governance. Detailed FSMS plans should be developed for each In-house Canteen as the company itself is responsible for Food Safety in this case and its employees would be at legal risk in case of any mishaps.

- Vendors shall get Licensing for themselves and prepare their own FSMS Plan, but validation of the FSMS plan as per the Company's Standards is must, to protect the people and to safeguard the goodwill of the Company.

- Procurement of FSSAI License for the Central Kitchen of the Food Vendors is mandatory.

- Licensing/Registration of all the sites where food is being served to masses is mandatory. The Licenses/Registrations for these sites are either to be obtained by the Company or the Vendor, depending upon who is managing the operations at the site. If a canteen is being maintained and operated by the Company itself, the onus of Licensing/ Registration for the site would be with the Company.

- 1 person from the Company should be in charge of looking into Food Safety. He should be directly maintaining Food Safety measures in the Canteen and supervise the Food Safety initiatives at the Vendor's Kitchen. 1 person from each of the Food Vendors should be nominated as Responsible Person for Food Safety, carrying the daily internal audits, facilitating periodic external audits and co-ordination with the responsible person from the company.

- Monthly/Quarterly/Half Yearly Action plans and schedules are to be prepared for activities to ensure adherence to the FSMS Plans developed.

- Checklists with Adequate and Relevant points are to be prepared for conducting the Daily Internal Audits. The Checklist should be objective, based on points of Visual Confirmation and should cover all the relevant areas of Food Safety. Help of an external professional should be sought if required.

- Whenever a Non Compliance is found in the Internal Audits, the External Audits or the Sample Test Reports, NCR should be filed.

- NCR should always be filed along with mention of the necessary corrective actions and timelines for implementation of the Corrective Actions. Once the Corrective Actions are implemented, the NCR is to be closed by attaching the required proof of Corrective Action Implementation.

Processes & Service Level Agreement

Food Sampling & Laboratory tests

- The Service Provider will retain food samples of all the portions/items served, each day for showing to various competent parties as necessary. The food samples will be held in a separate refrigerator and will be marked accordingly. The samples will be of a volume of 25 grams. per product, for a period of time of at least 48 hours from the date of the food distribution. The Service Provider will prepare a special form for recording the food samples kept each day, including: date on which cooked/preparation of the food, the quantity prepared and time of serving (date and hour)

- The samples must include all the food variety, which will be serving to Company employees.

- The Service Provider will, once every three months, send samples of food for laboratory examination on the account of the Service Provider (the examination will be carried out without prior notification and on a random basis by a representative of the Company). The Service Provider undertakes to forward to client (company) the examination reports of the laboratory. Without prejudice to the generality of the aforesaid, the company retains the right to demand that the Service Provider sent additional food samples for laboratory examination on the account of the Service Provider, this being without having to explain its demand.

- The Service Provider will present the results to a Company representative.

- The Company is entitled in any time to instruct the Service Provider to send the reserved food sampling to a laboratory, as far as required.

- All examinations will be made by the Service Provider and at his expense.

Certification & License

- Service Provider will present all the required approvals and licenses according to the local authorities in order to active the kitchen he owns. i.e. Shop Act, FDA (food & drug administration), FSSA (food safety & standards act) or FSSAI (food safety & standards authority of India) whichever is applicable to the vendors establishment

- The Service Provider will cooperate with the Company if needed, in order to get approvals and licenses to active the dining hall in the Company structure.

Delegate Manager of the Catering Company

- The Service Provider will employ a professional manager with a relevant certificate for of the operations of the cafeteria and pantry services, from a catering/hotel background.

- The manager on behalf of the Service Provider will be in the cafeteria whenever it's working and will supervise on the Service Provider's full commitment existence, according to this document's requirement.

- The manager will be the only contact person in front of the Company's focal points at site and will answer all their requirements.

Chefs

- The Service Provider will employ a professional Executive Chef with Degree in Hotel Management and minimum 5 years of experience in similar Corporate Catering Services at the similar hierarchy/level with cooking knowledge in depth of Indian & continental cuisine.

- The Service Provider will employ Chou Chefs and Head Cooks of any product line should be well qualified, having at least a Diploma in Hotel Management or relevant minimum 5-7 years' experience.

- The Competency levels of the Chefs can be vouched for by the Company, to ensure the consistency and taste of the Product/food preparation

Safety Guidelines

- Defined quality checks need to be carried out from procurement of raw material till pick-up of food, to ensure quality of food and its hygiene.

- The Central Kitchen should be meticulously cleaned and maintained at all times. The housekeeping activity is to be aligned with the food preparation cycle so as to provide a clean environment, prior to, during and after food preparation.

- Raw material purchased for food preparation should be from established & known Service Providers, should be fresh in appearance, free from physical impurities/fungal growth. Packaged food products should have the pack seals intact/without holes/leakage/dents/ puffing & rusting signs.

- The storage of raw material is to be carried out in hygienic manner and perishables must be stored according to the recommended temperatures. Storage area should be clean & free from debris, empty boxes/other refuse, well ventilated & well lighted, free from insects, pests or their remains. All foods & paper/packaging supplies should be stored off the floor and away from the walls (at least 6 inches)

- Raw material used for food preparation should be used by following first-in-first out (FIFO) method and products should be labeled with name & date (expiry/delivery)

- Before usage of raw material, it should be ensured that the raw material is not out of expiry.

- Only potable water from safe source is used for preparation of raw materials/ or as an ingredient in food products. Also potable water is used for washing of food products.

- The food components must be cleaned and inspected prior to their use for any preparation. Any foreign/ unwanted material must be removed so as to provide any ill-effects on the product.

- Testing of food will be carried out, before it is dispatched to the client site.

- All food containers must be kept clean & dry prior to filling for dispatch.

- The vehicle carrying the packed food must be kept clean at all times and free from pests

- The loading of the vehicle will be carried out in a segregated area, free from housefly nuisance and separated by air curtains. Containers to be of thermal quality & air tight.

- On completion of loading the food delivery vehicle will be locked and the same is not to stop at any other destination than the intender

- At receiving point, care must be taken to unload the food without disturbing its contents or roughly handled, so as to cause damage to them.

- Vehicle should always reach in locked condition and containers are sealed & air tight.

- While unloading & taking food containers to the Cafeteria, follow correct posture for unloading, shifting and loading of containers.

- In case Service Elevator is used, specific safety instructions related to use the lifts must be adhered to all the time.

- Adequate staff should be available to carry the food containers to ensure load is distributed equally with normal lifting capacity.

- All the personnel working at site must be literate and regularly trained for using safe practices in all their work.

- First Aid Box should be available in the Cafeterias

- Proper tools (peelers, knives, bottle openers, etc.) should be provided for any activity being carried out. All such tools (knives) are stored in specified drawers.

- While cleaning the tools, they should be cleaned from the blunt side to avoid any cuts.

- Rubber soled Boots/Footwear's should be used by staff working on slippery floors.

- Regular mopping & drying of the floor is to be carried out.

- Rubber mats are to be provided on the floor which are oily/ where Dish Cleaning Activity is carried out, so as to provide safety.

- While handling any equipment, the safety instructions provided by the manufacturer of that equipment should be strictly followed.

- Transfer of food from containers to Ben-Marie should be done only by using ladles.

- Unconsumed confectionaries to be returned back to Vendor on daily basis

- Bakery items are kept in refrigeration system which are maintained at advisable temperature

- Items are kept for sale/ usage must be consumed well within the stipulated expiry date. Storage of such material close to expiry date will be avoided at all times. This is to be followed meticulously in case of perpetually used material like bread (48 hrs. expiry).

- The refrigeration system should be calibrated on a regular basis

- Plates should be cleaned in the Dish Washing Machine only and staff should be wearing proper apparatus.

- While manual cleaning, proper care to be taken, to ensure that there are not cuts because of sharp edges.

- Water in Ben-Marie is to be drained once in two days (daily if any food particles have been dropped). This water is to be heated not in excess of 70deg while commissioning the use for the day.

- Experienced staff to do cutting as well as washing of knives & forks.

- Safety distances of the gadgets, emitting heat is to be maintained from the wall end, so as to ensure that the industrial switches do not get overheated.

- Any equipment installed at a height (fly catcher) should be cleaned only after it is unplugged. Also stable ladders are to be used for such purposes.

- Whenever the floor is being cleaned with water, "Wet floor" signage needs to be put, so that everyone is aware of that.

- On completion of activity at the Cafeteria, all waste food will be accumulated and accounted for to record the generation on a day to day basis.

- Such accumulated wastage will be passed on to authorities as per the collection schedule.

- At the end of the day's activity, the entire cafeteria will be cleaned and kept dry so as to avoid fungus/ germination activity.

- On weekends the cafeteria would be cleaned in depth to cover all crevices, corners, toilet blocks, cob webs, etc.

Fines & Penalties

Sr. No	Fines/ Penalties:	Fine Amount (Per Incident)
1	Working without a head covering	Xyz amount (as agreed mutually considering the severity of the offence)
2	Working without gloves	
3	Serving recycled food that is forbidden to be recycled	
4	Serving hot food at a temperature of less than 68 degrees	
5	Serving cold food (salads) at a temperature exceeding 8 degrees	
6	Transporting food in conditions contrary to the regulations for the transporting of food, as defined in this Agreement	
7	Kitchen not clean and/or kitchen equipment not clean/ not in working condition	
8	Failure to display an updated business licenses certificate in the premises	
9	Use of and frying in oil that is not clean (cloudy)	
10	Shortages of food, utensils, cutlery and crockery, condiment sets, napkin holders and so on during the operating hours of the cafeteria	
11	Shortage of delegated staff at site during operations	
12	Any foreign body/ insect, worms/hair/stones found in food	
13	Meat, fish, poultry products not sourced from a licensed vendor	
14	Sub-standard food quality/ texture/ taste	
15	Dining tables remaining un cleaned for more than 5 minutes	
16	Corrective Suggestions from the company personnel/ Outside Auditors not implemented within the agreed schedule/one month, whichever is earlier	

Challenges & Mitigations

Challenges	Mitigation
Monotony in food variety & taste	Vendor Chef must be able to provide with new inclusions as variety & different cuisine for change of taste from time to time

Adherence to café agreement SLA/ Labor staff attrition - Inadequate manpower considering size of site & its services requirement	All of the said points must be considered strictly as agreed in the contract agreement & the vendor must be held responsible/ accountable to provide us with improvement initiatives with action plans
Compliances to legal requirements like - audits, sub vendor agreements in place - e.g. kitchen equipment's AMC, Insurances etc..	
Food quality & consistency	
Personal & food hygiene	
Premises maintenance	

Role of a cafeteria manager

♦ Finalize monthly menus with the caterer vendor & food committee.

♦ Annual audits for kitchen equipment's health and AMC status.

♦ Ensure smooth serving of food in cafeteria, general upkeep (cleanliness) of all Cafeterias.

♦ Prepare food distribution % for services in Pantries/ kitchenettes as per floor/ occupancy.

♦ Food/Snacks testing on daily basis & communicating to the vendors about Quality & Quantity short falls. Keeping record of checks done on daily basis.

♦ Responsible for introducing new variety from time to time to avoid monotony.

♦ Preparing critical analysis on café costs/ food related issues/complaints etc.

♦ Conducting regular meetings with Cafeteria vendor's management to address issues related to Cafeteria food, Snacks, Fruits & 24 x 7 Counter.

♦ Regular surprise visits to Central/ Mother kitchen to ensure quality monitoring mechanism is in place & to check that food is prepared in hygienic condition.

♦ To ensure ISO/ EHS practices are followed in the Central/ Mother kitchen/ Cafeteria.

♦ Handling requirements of special menu & dining arrangements for VIP's & Guest's and on special company events/ occasions.

♦ Maintaining café inventory of equipment's/ cutlery, crockery/ tables & chairs & prepare periodic repair list

♦ Maintain & update café MIS

SUSTAINABILITY

Because of the industrial revolution in 18th to 19th centuries, humans started tapping into the natural resources of energy i.e. fossil fuels. Fossil fuels are an essential commodity to power engines and machineries. Also due to advances in medicine had decreased the death rates by improving health conditions from disease.

In the mid-20th century, an environmental movement pointed out that there were environmental costs associated with the utilization of minerals and fossil fuels. In the late 20th century, environmental problems became global in scale. The 1973 and 1979 energy crises demonstrated the extent to which the global community had become dependent on non-renewable energy resources.

In the 21st century, there is increasing global awareness of the threat posed by the human greenhouse effect, produced largely by forest clearing and the burning of fossil fuels.

Achieving sustainability will enable the Earth to continue supporting human life.

Sustainability means that without affecting environment and natural resources how a human race can survive and develop. The main factor for sustainability is sustainable development which contains four interconnected domains:

1. Ecology
2. Economics
3. Politics
4. Culture

A healthy ecosystem and environments are very important for survival of humans and other living beings. There are various methods to avoid negative impact on environment, which is as appended below:

1. Environmental-friendly chemical engineering
2. Environmental resources management and environmental protection.
3. Green energy and reduction in greenhouse gases.

In order to achieve sustainability, governments should play major roles with respect to drafting and implementation of policies and they should work towards urban planning and transport. However this is not only responsibility of government for sustainable development, it is responsibility of all individual and every citizen are required to change and control their lifestyle and ethical consumerism. There are various ways to live sustainably which can be:

1. Reorganization of villages and cities, by transforming them into ecologically balanced villages and municipalities and making sustainable cities. Example government should promote Private Township with integrated sustainable and zero discharge arrangements in place.

2. Restructure the economic sectors by using permaculture and sustainable agriculture. Constructing and maintain green buildings and using sustainable architecture.

3. Conduct research for development of new technologies such as green technologies, renewable energy and sustainable fission and fusion power. Reduce the cost of production of sustainable energy systems.

4. Design systems in such a manner that they are flexible and reversible.

5. Adjusting individual lifestyles that conserve natural resources.

In spite of human race is aware of fast depleting natural resources, environment changes and global warming it is still doubtful that human societies will achieve environmental sustainability. The reason behind the same is non-participation of citizens and poor actions by governments.

Three Important sustainability factors

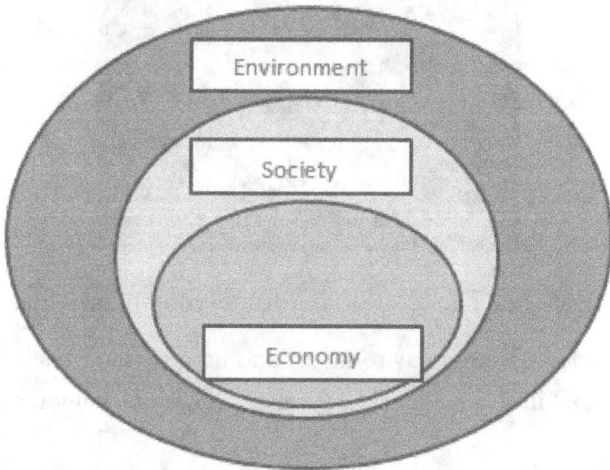

The above diagram indicates that the economy & society are constrained by environmental limit.

The World Summit on Social Development conducted in 2005 had identified sustainable development goals, such as economic development, social development and environmental protection.

Some sustainability experts and practitioners have illustrated four important factors of sustainability that are environment, Economy, Society and Future generations. This aspect emphasizes on the long-term thinking associated with sustainability.

Sustainable development consists of balancing local and global efforts to meet basic human needs without destroying or degrading the natural environment.

In simple words sustainability is something that improves "the quality of human life while living within the carrying capacity of supporting eco-systems", though vague, conveys the idea of sustainability having quantifiable limits.

However sustainability needs a lot of actions instead of talks. It needs political & social will and proactive decision-making and innovation that minimizes negative impact and maintains balance between ecological resilience, economic prosperity, political justice and cultural vibrancy to ensure a desirable planet for all species now and in the future. The Earth Charter speaks of "a sustainable global society founded on respect for nature, universal human rights, economic justice, and a culture of peace." This suggested a more complex figure of sustainability, which included the importance of the domain of 'politics'.

Principles of Sustainability

The principal of sustainability connects with many different disciplines and fields; in recent years this has been called as sustainability science.

As per United Nations the world needs to go for sustainable development which should include economic development, social development and environmental protection. The sustainable development has four domains of economic, ecological, political and cultural sustainability.

Sustainability of mother earth totally depends on economic sectors, ecosystems, countries, municipalities, neighborhoods, home gardens, individual lives, individual goods and services, occupations, lifestyles, behavior patterns and so on.

Consumption

Humans have started destruction of the Earth's ecosystems, by over utilizing the natural resources. The environmental impact totally depends on the human population and on utilization of resources, whether or not those resources are renewable, and the scale of the human activity relative to the carrying capacity of the ecosystems involved.

To understand human impact on environment mathematically a formula was developed in 1970s and is called the I PAT formula. This formulation attempts to explain human consumption in terms of three components: population numbers, levels of consumption (which it terms "affluence", although the usage is different), and impact per unit of resource use (which is termed "technology", because this impact depends on the technology used). The equation is expressed:

$$I = P \times A \times T$$

Where: I = Environmental impact, P = Population, A = Affluence, T = Technology

Population

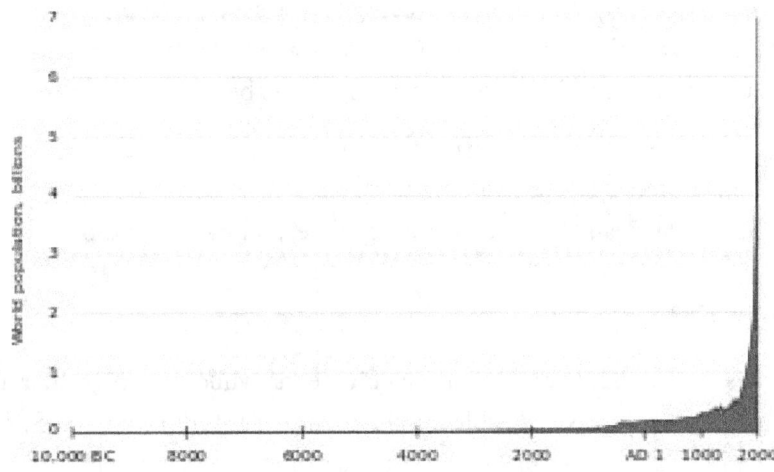

The above graph shows human population growth from 10,000 BC – 2000 AD, illustrating current exponential growth. It shows that there was a population growth explosion after 19th century.

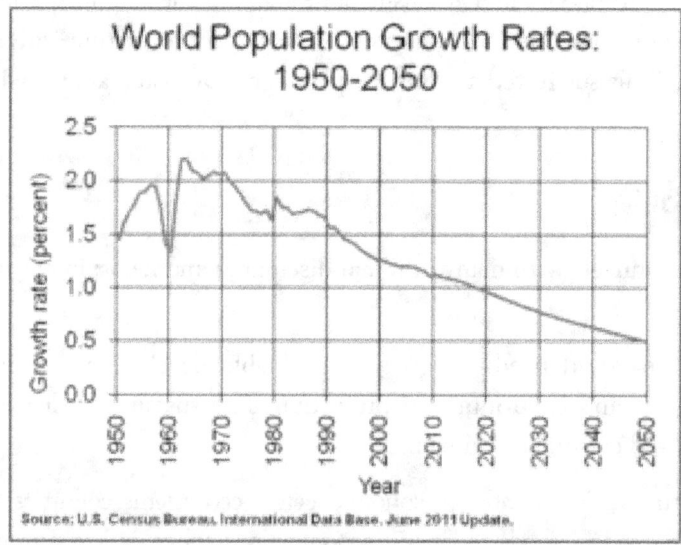

World population growth rate, 1950-2050, as estimated in 2011 by the U.S. Census Bureau, International Data Base

As per the United Nations population estimates and projections, the world population is projected to be 7.4 billion by March 2016, and expected to reach 9.5 billion in 2050, which is up from estimated 7.3 billion.

Carrying capacity

The researches have indicated that the humans are living beyond the carrying capacity of planet Earth and that this cannot continue for a longer time. A 2012 review in Nature by 22 international researchers expressed concerns that the Earth may be "approaching a state shift" in its biosphere.

The Ecological footprint means human consumption in terms of the biologically productive land required to full fill human requirement with respect to food, housing and waste disposal. It is considered that 2.7 hectares of land is required per person, which is 30% more than the natural biological capacity of 2.1 hectares per person. In this case there will be an ecological deficit, which actually today's generation is forcibly using which belong to future generation.

Human impact on biodiversity

Human impacts on the Earth are clearly visible on natural resources such as water, oxygen, soil, carbon, nitrogen and phosphorus. These impacts are now visible and the whole world is going through tough time. if human behavior is not corrected, the days are not far when there will be war and rise in crime because of scarcity of these resources.

The Millennium Ecosystem Assessment report refers to natural systems as humanity's "life-support system", providing essential "ecosystem services". The assessment measures 24 ecosystem services concluding that only four have shown improvement over the last 50 years, 15 are in serious decline, and five are in a precarious condition.

Sustainable development

Sustainable development means fulfilling of human requirements while ensuring that the ecological balance is maintained. While the modern concept of sustainable development is derived most strongly from the 1987 Brundtland Report, it is rooted in earlier ideas about sustainable forest management and twentieth century environmental concern.

Environmental, economic and social well-being for today and tomorrow

"Sustainable development is development that meets the needs of the present without compromising the ability of future generations to meet their own needs. It contains within it two key concepts:

- the concept of needs, in particular the essential needs of the world's poor, to which overriding priority should be given; and

- the idea of limitations imposed by the state of technology and social organization on the environment's ability to meet present and future needs."

If you consider world as a system you will come to know that air pollution from once country can affect air quality of other country which may be remotely located. You also start to realize that the decisions our grandparents made about methods of farming is continuing to affect agricultural practice today and also today's economic policies will be having an impact on urban poverty when our children are adults.

Ecology

The ecological sustainability is availability of quality air, water, food and shelter. These are important factors for a sustainable development. The ecology sustainability can be achieved by addressing public health risk through investments in ecosystem services can be a powerful and transformative force for sustainable development.

Environment

Natural resources are such as the todays status of air, water, and the climate are of major concern. Environmental sustainability wants society to design their activities in such a way that human needs are fulfilled by preserving the life support systems of the planet. E.g. water sustainably can be achieved by charging the ground water, using rain water harvesting system, recycling of water etc. etc.

Average consumption of renewable resources	State of environment	Sustainability
More than nature's ability to replenish	Environmental degradation	Not sustainable
Equal to nature's ability to replenish	Environmental equilibrium	Steady state economy
Less than nature's ability to replenish	Environmental renewal	Environmentally sustainable

Agriculture

We all know that today we are excessively using chemical fertilizers for farming. This in turn is degrading the quality of soil & also has ill effect on the human health. For a sustainable agriculture humans are required to use environmentally friendly methods of farming. This in turn will prevent adverse effects to soil, water, biodiversity, surrounding or downstream resources. Elements of sustainable agriculture include permaculture, agroforestry, mixed farming, multiple cropping, and crop rotation, usage of vermiculture instead of chemical fertilizers.

Energy

The energy generated using natural resources such as wind & solar are known as Sustainable energy or also as green energy. The green energy is clean and doesn't damage the environment. The most common types of renewable energy are solar and wind energy.

India has a great opportunity to use solar energy as it has 300 clear days in a year with sun lights. It is estimated that India can generate up to 5,000 trillion kilowatt-hours (kWh) per year through solar energy, which exceeds the possible energy output of all fossil fuel energy reserves in India.

The daily average solar power plant generation capacity over India is 0.25 kWh per m2 of used land area.

As of 31 August 2015, the installed grid connected solar power capacity is 4,229.36 MW and India expects to install an additional 10,000 MW by 2017 and a total of 100,000 MW by 2022. Indian government is targeting an investment of US$100 billion and 100 GW of solar capacity by 2022.

The toxicants from fossil fuel base power plants are major contributors to health problems in the communities. In the long run, sustainable development in the field of energy is also deemed to contribute to economic sustainability and national security of communities, thus being increasingly encouraged through investment policies by all the governments across the globe.

A sewage treatment plant that uses solar energy, located at Santuari de Lluc monastery, Majorca.

Transportation

One third of total emission of greenhouse gases is generated by transport. Government has to take an immediate initiative to bring down emission levels. Some of the western countries have taken initiative to make their transportation more sustainable in terms of long-term and short-term. There are many ways in which the transportation can be improved. The government has to ensure that their public transport system is modern and efficient, they can us metro trains, bus services etc. etc. They can also provide exclusive cycling and walking pathways.

Due to lack of public transport, people prefer to travel by their own means of transport such as car, scooter or bikes. The government has to come up with some plans to reduce the total number of vehicle trips in order to lower greenhouse gases emission. Such as:

- Improve public transit through the provision of larger coverage area in order to provide more mobility and accessibility, new technology to provide a more reliable and responsive public transportation network.

- Encourage walking and biking through the provision of wider pedestrian pathway, bike share station in commercial downtown, locate parking lot far from the shopping center, limit on street parking, and slower traffic lane in downtown area.

- Increase the cost of car ownership and gas taxes through increased parking fees and tolls, encouraging people to drive more fuel efficient vehicles. They can produce social equity problem, since lower people usually drive older vehicles with lower fuel efficiency. Government can use the extra revenue collected from taxes and tolls to improve the public transportation and benefit the poor community.

- Encourage corporates to provide mass transport facility for their employees.

- Discourage employees from using personal vehicle.

- Encourage car pooling's

Economics

There is lot of poverty in rural areas and environmental resources are being overexploited in almost all parts of the world. The rural and environmental resources should be treated as important economic assets as these are our natural capital. Economic development has traditionally required a growth in the gross domestic product. Sustainable development will help in improvement in the quality of life for all, but at the same time this will also require a decrease in resource consumption. According to ecological economist Malte Faber, ecological economics is defined by its focus on nature, justice, and time.

A World Bank study from 1999 concluded that based on the theory of genuine savings, policymakers have many possible interventions to increase sustainability, in macroeconomics or purely environmental. A study from 2001 noted that efficient policies for renewable energy and pollution are compatible with increasing human welfare, eventually reaching a golden-rule steady state. The study, Interpreting Sustainability in Economic Terms, found three pillars of sustainable development, interlinkage, intergenerational equity, and dynamic efficiency.

A meta review in 2002 looked at environmental and economic valuations and found a lack of "sustainability policies". A study in 2004 asked if we consume too much. A study concluded in 2007 that knowledge, manufactured and human capital (health and education) has not compensated for the degradation of natural capital in many parts of the world. It has been suggested that intergenerational equity can be incorporated into a sustainable development and decision making, as has become common in economic valuations of climate economics. A meta review in 2009 identified conditions for a strong case to act on climate change, and called for more work to fully account of the relevant economics and how it affects human welfare. According to John Baden "the improvement of environment quality depends on the market economy and the existence of legitimate and protected property rights." They enable the effective practice of personal responsibility and the development of mechanisms to protect the environment. The State can in this context "create conditions which encourage the people to save the environment."

Business

The corporate has to take responsibility for sustainability and make efficient use of natural capital. This eco-efficiency is usually calculated as the economic value added by a firm in relation to its aggregated ecological impact. This idea has been popularized by the World Business Council for Sustainable Development (WBCSD) under the following definition: "Eco-efficiency is achieved by the delivery of competitively priced goods and services that satisfy human needs and bring quality of life, while progressively reducing ecological impacts and resource intensity throughout the life-cycle to a level at least in line with the earth's carrying capacity."

Similar to the eco-efficiency concept but so far less explored is the second criterion for corporate sustainability. Socio-efficiency describes the relation between a firm's value added and its social impact. Whereas, it can be assumed that most corporate impacts on the environment are negative (apart from rare exceptions such as the planting of trees) this is not true for social impacts. These can be either positive (e.g. corporate giving, creation of employment) or negative (e.g. work accidents, mobbing of employees, human rights abuses). Depending on the type of impact socio-efficiency thus either tries to minimize negative social impacts (i.e. accidents per value added) or maximize positive social impacts (i.e. donations per value added) in relation to the value added.

Architecture

Architecture needs to promote a sustainable approach towards construction. Architects should be focused towards smart growth, architectural tradition and classical design. The sustainable architecture shall ensure the less commuting distance between work area and living area. Also construction of green and energy efficient buildings will help in sustainability.

Reference: International Institute for Sustainable Development.

What a Facility Manager can contribute towards Sustainability

1. Ensure that highly toxic chemicals are avoided for cleaning purpose.

2. Use eco-friendly cleaning chemicals.

3. Use bio blocks in urinals, instead of using water for flushing.

4. Use water saving fittings in taps which saves water.

5. Avoid usage of plastic products & packaging

6. Avoid installation of carpet in office area if mandatory by policy use low VOC carpet

7. Avoid synthetic paint in your office or building, use low VOC paints.

8. Use biochemical to dissolve fats collected in oil water separator.

9. Avoid using chemical fertilizer. Use vermiculture or cow dung as manicure/fertilizer in your green area.

10. Arrange for rain water harvesting in your facility. Use this system to charge underground water.

11. Make use of recycled water for flushing of toilets, urinals & watering of gardens.

12. Make use of kitchen water for gardening purpose.

13. Avoid usage of paper napkins. And make use of both sides of the paper for printing.

Note: For production of 1 ton of paper you need 50,000 ltr of water. It means if we consider 1 tissue paper weighs 1 gram then you need to spend 55ltr of water to recycle that paper napkin + chemicals such as caustic soda etc. + electricity.

14. Use public transport or provide company employees with company transport this will avoid usage of personal cars.

15. Ask your transport agency to supply CNG fitted vehicles only.

16. Use energy efficient lighting fixtures such as LED lights.

17. Use BEE star rated electrical or electronic equipment's. Try and procure 5 Star rated equipment's only.

18. Use motion sensors for washrooms, meeting & conference rooms and senior employee's cabin.

19. Use lux level sensors for street light controls and office areas near the façade.

20. For construction purpose make use of fly ash bricks.

21. When looking for new office spaces try and look for Green Certified buildings only.

22. Try and utilize renewable energy such as solar or wind energy. If you have an independent building you can install solar power plant on the terrace, now a days agencies are available who can install this plant for your company without any Capex.

ELECTRICITY DISTRIBUTION

There are two types of electricity distribution i.e Direct Current (DC) and Alternate Current (AC). Today's electricity supply which we receive in our office and at home is purely AC supply. The advantage of having an AC supply is that it needs less infrastructure and is easy to transfer. However it is to be handled carefully, if not it may lead to serious injuries and accidents such as fire due to short-circuit which further may cause loss of life, property and closure of business. Let's understand how and electricity is distributed and how it is made safe to use it.

Power distribution is the crucial link and the weakest in the electricity supply chain. The distribution segment starts from electricity generation plant at the 66/33 kV level. The standard voltages on the distribution side are, 66 kV, 33 kV, 22 kV, 11 kV and 0.4kV/o.230 kV, besides 6.6 kV, 3.3 kV and 2.2 kV. Depending upon the quantum of power and the distance, lines of appropriate voltages are laid.

The main distribution equipment comprises HT and LT lines, transformers, substations, switchgears, capacitors, conductors and meters. HT lines supply electricity to industrial consumers while LT lines carry it to residential and commercial consumers.

In India the type of electricity distribution is Radial and permissible limit of voltage variations allowed in the distribution system is +- 6%.

Electricity distribution system is as follows:

1. Power Generation Plant
2. Central Station
3. Transmission Station
4. Distribution stations
5. Feeder Pillars
6. Consumers (Industrial, Commercial, Residential)

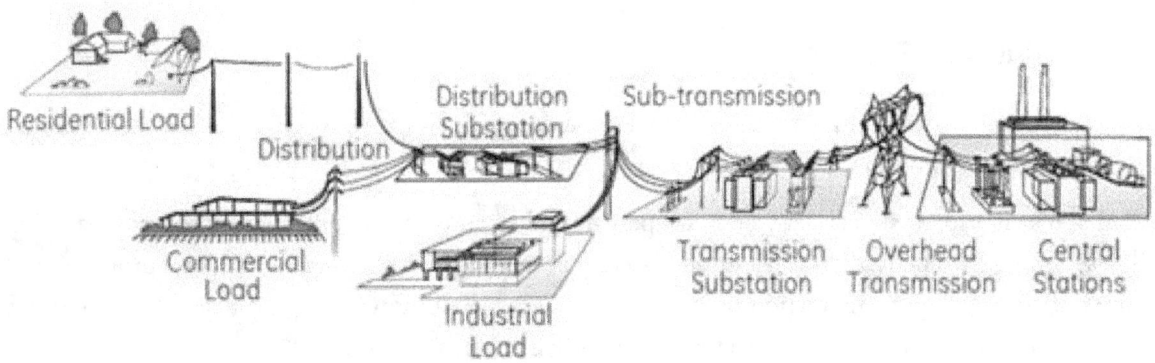

Distribution System Layout

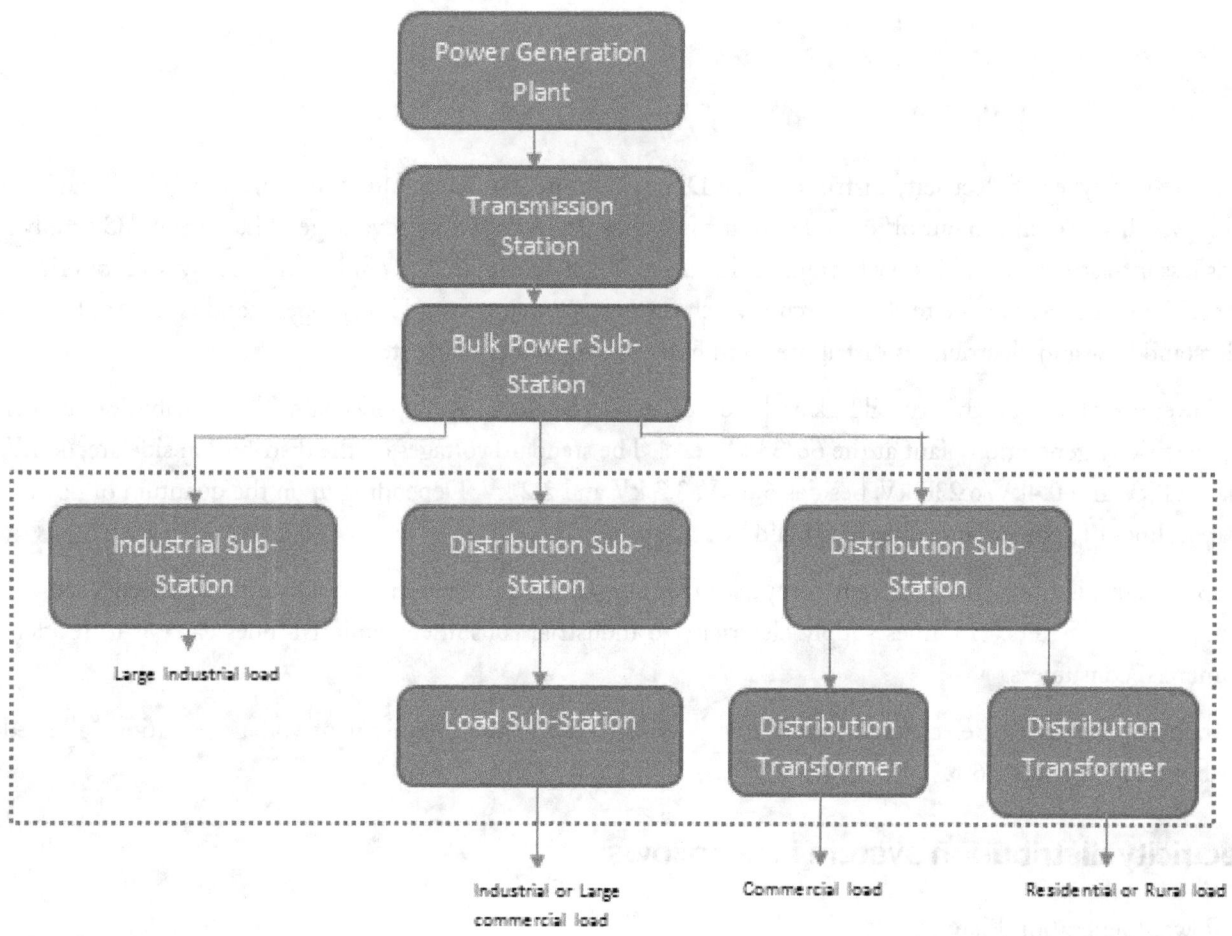

Power generation is normally takes place at a voltage between 3.3kV to 22kV which is a medium voltage. The voltage is then stepped up-to a level of 110kV or 220kV (high Voltage) or even can be raised up to 400kV (extra high voltage) depending on the amount of power to be transmitted.

At the distribution station the power is stepped down to a voltage of 430/250V for end consumers considering the voltage drop in distribution lines while distributing. In India, the electricity supply to residential premises is at 240V, single phase at 50 Hz frequency and for commercial consumers it is supplied in three phase at 415V and 50 Hz frequency.

Types of Distribution System

There are three types of distribution system; they are Radial, Loop & Network Systems. There can be combination of these there systems for distribution.

The Radial distribution system is easy and cheapest to build and is widely used in sparsely populated areas and it is normally used in India. The Radial distribution system will have only one power source for a group of customers. A power failure due to stoppage of generation on due to cable damage will interrupt power in entire line. However now government & corporates are taking initiative for a Loop or Ring Main system.

A Loop System; loops through the service area and returns to the point of origin. The loop system is also known as RMU (Ring Main System). The loop is usually tied into an alternate power source. By placing switches in strategic locations, the utility can supply power to the customer from either side.

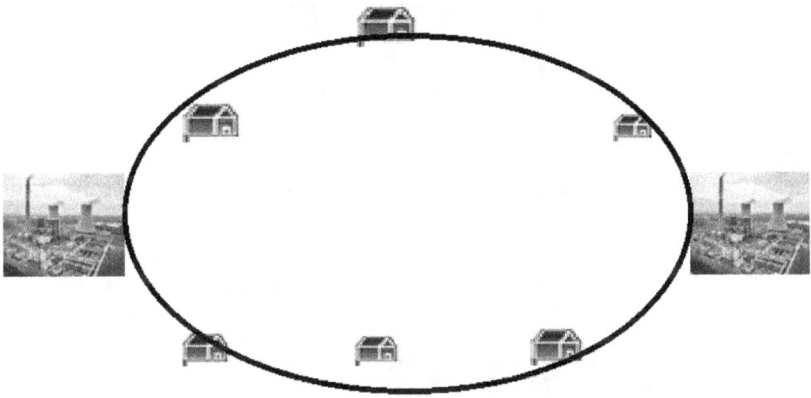

In Loop system if one source of power fails, the power breaker can operate manually or automatically, and power can be fed to customers from the other source, without any delay or interruption.

The loop system provides better continuity of service than the radial system, with only short interruptions for switching. In the event of power failures due to faults on the line, the utility has only to find the fault and switch around it to restore service. The fault itself can then be repaired with a minimum of customer interruptions.

The loop system is more expensive than the radial because more switches and conductors are required, but the benefit of this system is far better than any other system.

Network systems are the most complicated and are interlocking loop systems. A given customer can be supplied from two, three, four, or more different power supplies. Obviously, the big advantage of such a system is added reliability. However, it is also the most expensive. For this reason it is usually used only in congested, high load density municipal or downtown areas.

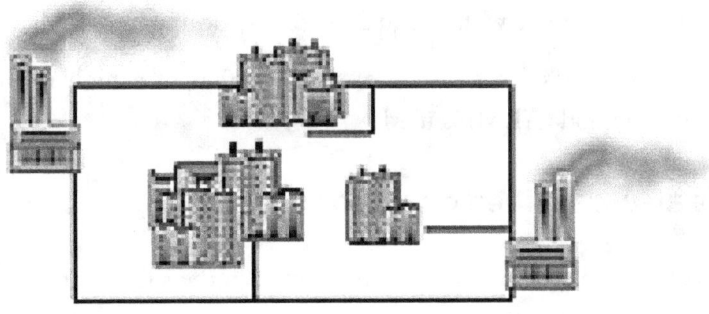

Ring Main Unit

RMU panel generally used in Loop or Ring Main system. It is generally gets supply from more than one feeder as incoming supply so that in case of failure of one feeder, power can be fed uninterruptedly from other feeders to the consumers.

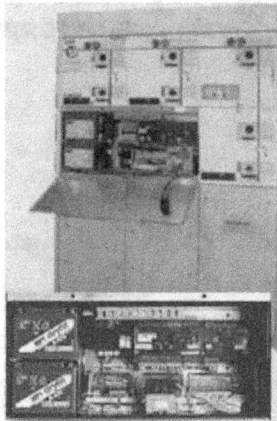

RMU is used only on HT side and it has three breaker. It has two inputs and one outgoings and is primarily used for power feeding purposes. It also helps in protecting secondary side of transformer from occasional transient currents.

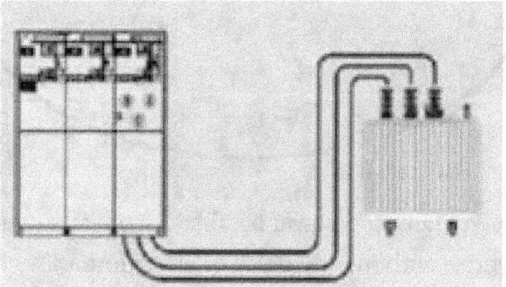

Faults in Distribution Lines

Transient Faults: A transient fault is a fault that is no longer present if power is disconnected for a short time. For eg. Momentary due to tree or bird contact.

Persistent Faults: A persistent fault does not disappear when power is disconnected. Faults in underground power cables are often persistent. These occur due to damage because of cable fault or digging.

Fault can further be classified; as symmetric or asymmetric faults.

Symmetric faults: Equal fault current in line with 120 deg displacement. Rare in nature.

Asymmetric faults: Unequal fault currents with unequal phase displacement. Most likely to occur.

Fault Detection in Distribution Lines: To detect the exact location of a high resistance fault in underground transmission lines Arc Reflection Method (ARM) is used.

Distribution System is also classified on voltage

♦   Primary Distribution- The part of electrical supply system between the substation and distribution transformers is called the primary system.

• Secondary Distribution- The secondary distribution system receives power from the secondary side of distribution transformers at low voltage and supplies power to various connected loads via service lines.

Circuit Breakers

Electrical circuit breaker is a switching device which can be operated automatically or manually for protecting and controlling of electrical power system. They save system from overloading or short circuit damage or to prevent from arc while operating by auto tripping. They work smoothly as long as the appliances have sufficiently resistant and do not cause any over current or voltage.

The reasons for heating up the wires are too much charge flowing through the circuit or short circuiting or sudden connection of the hot end wire to the ground wire would heat up the wires, causing fire. The circuit breaker will prevent such situations; which simply cut off the remaining circuit.

There are various types of circuit breakers and they are:

1. Air Circuit Breaker

2. SF6 Circuit Breaker

3. Vacuum Circuit Breaker

4. Oil Circuit Breaker

Air Circuit Breaker

Air Circuit Breaker is used for a power supply up to 15 KV.

Air circuit breaker operates in the air. The arc generated while operation of breaker is quenched at atmospheric pressure. The two types of air circuit breakers are

• Plain air circuit breaker

• Air blast Circuit Breaker

Plain Air Circuit Breaker

Plain air circuit breaker is also known as Cross-Blast Circuit Breaker. In this, the circuit breaker is fitted with a chamber which basically surrounds the contacts. This chamber is known as arc chute.

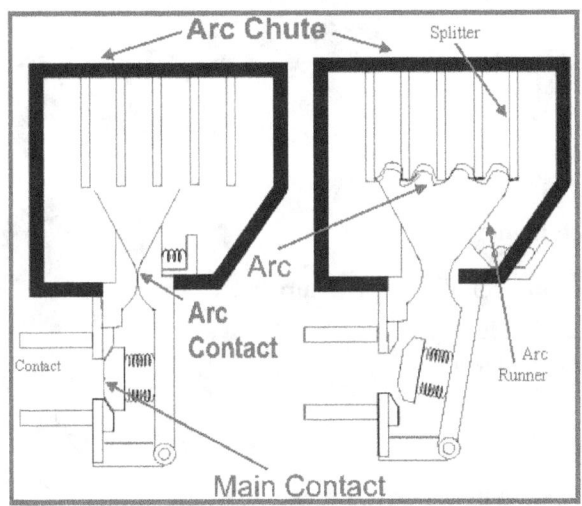

While operation the arc is made to drive in this arc chute and this arc chute helps in cooling of the air circuit breaker. The internal walls of arc chute are shaped in such a way that arc is not forced into close proximity. It will drive into the winding channel projected on an arc chute wall.

The arc chute will have many small compartments and has many divisions which are metallic separated plates. Here each of small compartments behaves as a mini arc chute and metallic separation plate acts like arc splitters. All arc voltages will be higher than the system voltage when the arc will split into a series of arcs. It is only preferable for low voltage application.

Air Blast Circuit Breaker: Air blast circuit breakers are used for system voltage of 245 KV, 420 KV and can be even more. Air blast circuit breakers are of two types:

- Axial blast breaker
- Axial blast with sliding moving contact

Axial Blast Breaker: In the axial blast breaker the moving contact of the breaker will be in contact. The nozzle orifice is a fixed to the contact of a breaker at a normal closed condition. To quench the arc a high-pressure air is introduced into the chamber. It ensures that voltage is sufficient to sustain high-pressure air when flowed through nozzle orifice.

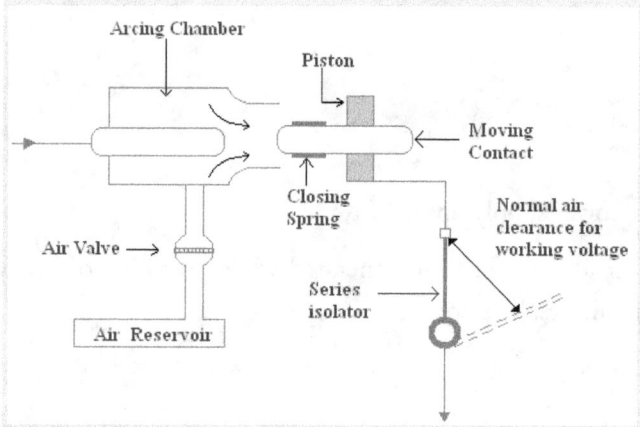

Advantage of Air Circuit Breaker

- It is used where frequent operation is required because of lesser arc energy.
- It is risk free from fire.
- Small in size.
- It requires less maintenance.
- Arc quenching is much faster
- Speed of circuit breaker is much higher.
- The time duration of the arc is same for all values of current.

Disadvantage of Air blast Circuit Breaker

- It requires additional maintenance.
- The air has relatively lower arc extinguishing properties

- It contains high capacity air compressor.

- From the air pipe junction there may be a chance of air pressure leakage

- There is the chance of a high rate rise of re-striking current and voltage chopping.

Application and Uses of Air Circuit Breaker

- It is used for protection of plants, electrical machines, transformers, capacitors and generators

- Air circuit breaker is also used in the Electricity sharing system and GND about 15Kv

- Also used in Low as well as High Currents and voltage applications.

SF6 Circuit Breaker

SF6 Circuit Breaker is used as HT Breaker for high voltage electrical power system from 33KV to 800KV. In the SF6 circuit breaker the current carrying contacts operate in sulphur hexafluoride gas. It has an excellent insulating property and high electro-negativity. It has a high affinity of absorbing free electron. The negative ion is formed when a free electron collides with the SF6 gas molecule and is absorbed by that gas molecule.

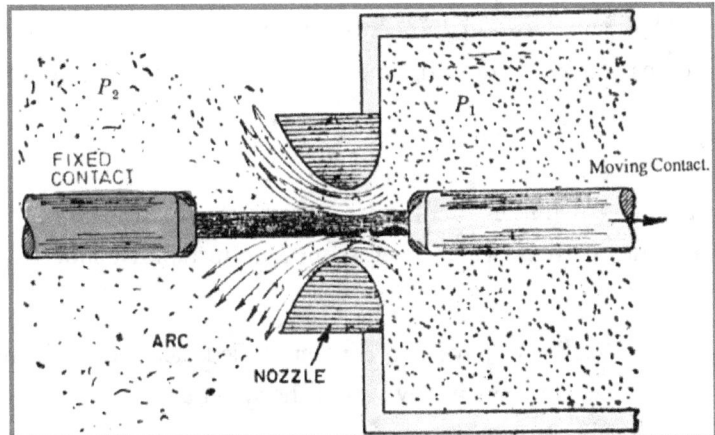

The negative ions which are formed will be much heavier than a free electron. Therefore, when compared with other common gases overall mobility of the charged particle in the SF6 gas is much less. The mobility of charged particles is majorly responsible for conducting current through a gas. Hence, for heavier and less mobile charged particles in SF6 gas, it acquires very high dielectric strength. This gas has a good heat transfer property because of low gaseous viscosity. SF6 is 100 times more effective in arc quenching media than air circuit breaker.

Types of SF6 Circuit Breaker

- Single interrupter SF6 circuit breaker applied up to 220V

- Two interrupter SF6 circuit breaker applied up to 400V

- Four interrupter SF6 circuit breaker applied up to 715V

Maintenance of SF 6 HT Breaker

The maintenance process and schedules are same for all oil circuit breaker, air circuit breaker, SF6 circuit breaker and vacuum circuit breaker. In addition to that in SF6 CB some extra care to be taken.

1. The SF6 breaker must be checked for gas leakage, if unwanted low gas pressure alarm operates. The leakage can be checked using gas leakage detector.

2. If the circuit breaker is provided with gradient capacitors, these must be checked for oil leakage monthly. If leakage found plug it.

3. Dew point of SF6 should be checked with the help of dew point meter or hygro meters in every 3 to 4 years interval.

4. On an annual basis, undertake visual inspection for anti-condensation cabinet heaters, general condition of control hardware and mechanism components within the breaker cabinet.

5. Every 05 years one should undertake testing of timing, contact resistance, control and auxiliary alarms, SF6 moisture and low gas density alarms. Mechanism maintenance is also recommended during this interval.

6. Every ten 10 or 12 years (depending on OEM) a "major" maintenance is to be undertaken. This includes 05 year maintenance and a visual inspection of the interrupter contacts and replacement of the disturbed seals and desiccant.

7. The CBM recommendations "dove-tail" into the TBM practices stated above, but consider mechanical operations on the breaker (2000-2500 operations) for inspection and maintenance. CBM recommendations also consider the amount (kA) of fault duty and number of fault clearing operations the breaker has been called upon to interrupt, either form installation or from the most recent breaker major maintenance or overhaul.

Vacuum Circuit Breaker

In a Vacuum circuit breaker, vacuum is used to extinct the arc. It has dielectric recovery character, excellent interruption and can interrupt the high frequency current, which results from arc instability, superimposed on the line frequency current.

In the principle of operation of VCB will have two contacts called electrodes will remain closed under normal operating conditions. Suppose when a fault occurs in any part of the system, then the trip coil of the circuit breaker gets energized and finally contact gets separated.

The moment contacts of the breaker are opened in vacuum, i.e. 10^{-7} to 10^{-5} Torr an arc is produced between the contacts by the ionization of metal vapours of contacts. Here the arc quickly gets extinguished, this happens because the electrons, metallic vapours and ions produced during arc, condense quickly on the surface of the CB contacts, resulting in quick recovery of dielectric strength.

Advantage of VCB's

- VCBs are reliable, compact and long life

- They can interrupt any fault current.

- There will be no fire hazards.

- No noise is produced

- It has higher dielectric strength.

- It requires less power for control operation.

Maintenance of VCB

Maintenance of VCB should be undertaken annually. The maintenance team should undertake dismantling of contactors and CT's and clean them before reinstallation and test for its operations.

Oil Circuit Breaker

In Oil Circuit Breaker the arc quenching media is oil. In this kind of breaker only mineral oil is to be used. The moving contact and fixed contact are immerged inside the insulating oil. When the separation of current takes place, then carrier contacts in the oil, the arc in circuit breaker is initialized at the moment of separation of contacts, and because of this arc in the oil is vaporized and decomposed in hydrogen gas and finally creates a hydrogen bubble around the arc.

This highly compressed gas bubble around and arc prevents re-striking of the arc after current reaches zero crossing of the cycle. The OCB is the oldest type of circuit breakers and no more in use.

Other kind of LT Circuit Breakers

Electrical circuit breakers are a switching device which can be activated automatically as well as manually to control and protect an electrical power system respectively. The LT circuit breakers are of various types based on special categories they have been subdivided into.

MCB (Miniature Circuit Breaker)

MCB is an electromechanical device which guards an electrical circuit from an over current, that may effect from short circuit, overload or imperfect design. MCB's has replaced the fuse which has its own drawbacks. MCB's once trips can be easily reset. An MCB can be simply rearranged and thus gives a better operational protection and greater handiness without incurring huge operating cost. The operating principle of MCB is simple.

In simple wordings, MCB is a switch which routinely turns off when the current flows through it and passes the maximum acceptable limit. MCB's are designed to guard against over current and overheating.

MCB is a replacement of rewirable switch-fuse units for low power domestic and industrial applications. The MCB is a package of three functions in one and it ensures protection against short circuit, overload and switching. Protection of overload by using a bimetallic strip & short circuit protection by solenoid.

MCB's are available in different pole versions such as single, double, triple pole & four poles with neutral poles if necessary. The normal current rating is ranges from 0.5-63 Amperes with a symmetrical short circuit breaking capacity of 3-10 KA, at a voltage level of 230 or 440V.

Characteristics of Miniature Circuit Breaker

- Rated current is not more than 100 amperes
- Normally, trip characteristics are not adjustable
- Thermal/thermal magnetic operation

MCCB (Moulded Case Circuit Breaker)

The MCCB is used to control electricity supply and is having short circuit and overload protection. MCCB is an electromechanical device, which guards electrical system from short circuit and over current (over load). They offer short circuit and over current protection for circuits ranges from 63 Amps-3000 Amps. The primary functions of MCCB is to give a means to manually open a circuit, automatically open a circuit under short circuit or overload conditions.

In case of tripping, unlike a fuse, this circuit breaker can be simply reset. It is safe in operations and easy to handle.

Characteristics of MCC

- MCCB can be used for rated current up to 1000 amperes
- Its trip current may be adjusted
- It operates on thermal magnetic

ELCB (Earth Leakage Circuit Breaker)

The ELCB is used to protect the electrical system from the electrical leakage. If a person gets an electric shock, then this circuit breaker cuts off the power at the time delay of 0.1 secs for protecting the personal safety and avoiding the gear from the circuit against short circuit and overload.

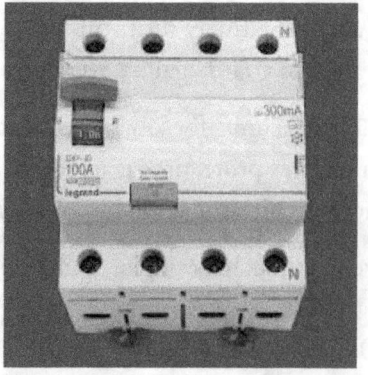

ELCB is a security device used in electrical system with high Earth impedance to avoid shock. This circuit breaker is a specialized kind of latching relay that has structures incoming mains power connected through its switching contacts so that this circuit breaker disconnects the power supply in an unsafe condition.

Characteristics of ELCB

♦   Phase (line), Neutral and Earth wire connected through ELCB.

♦   ELCB is working based on Earth leakage current.

♦   The safest limit of Current which Human Body can withstand is 30mA sec.

RCCB (Residual Current Circuit Breaker)

A RCCB is a current sensing equipment used to guard a low voltage circuit from the fault. It comprises of a switch device used to turn off the circuit when a fault occurs in the circuit. RCCB helps in guarding a person from the electrical shocks. This type of circuit breaker is used in situations where there is a sudden shock or fault happening in the circuit. It saves the infrastructure from personal injury and fire from short circuit.

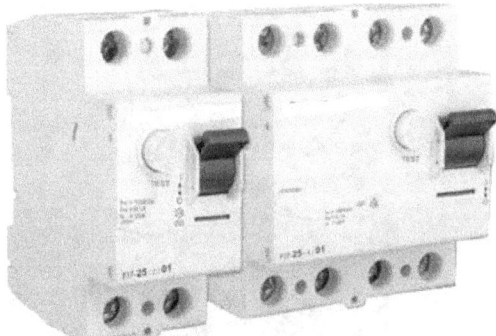

Characteristic of RCCB

♦   Both wires phase and neutral are connected through RCCB

♦   Whenever there is any ground fault occurs, then it trips the circuit

♦   The amount of current supplies through the line should go back through neutral

♦   These are a very effective type of shock protection

Difference between ELCB & RCCB

♦   ELCB is replaced with RCCB. ELCB is voltage-operated devices that are no longer available.

♦   RCCB is the new name that specifies current operated (hence, the new name to distinguish from voltage operated).

♦   The new RCCB is best because it will detect any earth fault. The voltage type only detects earth faults that flow back through the main earth wire so this is why they stopped being used.

♦   RCCB will only have the line and neutral connections.

♦   ELCB is working based on Earth leakage current. But RCCB is not having sensing or connectivity of Earth, because fundamentally Phase current is equal to the neutral current in single phase. That's why RCCB can trip when the both currents are different and it withstand up to both the currents are same. Both the neutral and phase currents are different that means current is flowing through the Earth.

◆ Function of both ELCB & RCCB are same, however their connectivity is different.

◆ RCCB does not necessarily require an earth connection itself (it monitors only the live and neutral).In addition it detects current flows to earth even in equipment without an earth of its own.

◆ This means that an RCCB will continue to give shock protection in equipment that has a faulty earth. It is these properties that have made the RCCB more popular than its rivals. For example, earth-leakage circuit breakers (ELCBs) were widely used about ten years ago. These devices measured the voltage on the earth conductor; if this voltage was not zero this indicated a current leakage to earth. The problem is that ELCBs need a sound earth connection, as does the equipment it protects. As a result, the use of ELCBs is no longer recommended.

Feeder Piller

A power box (USA) or feeder pillar (UK) is a cabinet for electrical distribution, mounted in the street and controlling the electrical supply to a number of consumers. It has bus bars in three phase plus an earthing bar. It receives electricity from the transformer and distribuites it to consumers.

HT Panel (High Tension Panel)

HT Panel is a high-tension panel, which receives supply from power grid or from power distribution company and supplies this power through Transformer for internal consumption. The power supply after step down to required voltage is distributed through synchronization/LT panel for end consumers.

Synchronizing Panel

Synchronizing Panel collects power supply from two or more different sources such as grid power and DG power and distributes it to the end consumers.

Synchronization means synchronization of frequency between two different power source and getting them on single frequency and connecting bot power source for further distribution. The frequency required to be maintained is +- 50 Hz.

It helps in transferring load from one unit to another as during service period, so that the unit requiring service can be easily shut off. In this way the critical load need not be interrupted and there is no production loss. During low load we can run any single unit, and synchronize more units as the load increases. This can be manual or automatic. The changeover time from one power source to another is not more than 60 seconds.

APFC (Automatic Power Factor Control) Panels

As per the law power factor of unity, i.e. .99 to 1 needs to be maintained. In case of non maintenance of power factor there is an arrangement for penalty to end consumers for not maintaining power factor. These panel is used in commercial buildings and industries where there is fluctuation in voltage and power supply. The electrical load required by a unit depends upon the type of machincrics, cooling plants and other devices installed. There is always a possibility of damage of these equipment's if power fluctuates. In case of fixed loads they can e safeguarded using capacitors, but in case of varied loads, a mechanism to switch in and switch out the capacitors is required which is basically handled using APFC panels.

Earthing

Earthing is used to connect metallic (conductive) Parts of an Electric appliance or installations to the earth (ground). In other words, to connect the metallic parts of electric machinery and devices to the earth plate or earth electrode (which is buried in the moisture earth) through a thick conductor wire (which has very low resistance) for safety purpose is known as Earthing or grounding.

Earthing can also be defined as the connection of the neutral point of a power supply system to the earth ensure minimize risk during discharge or leakage of electrical energy.

Importance of Earthing

The primary purpose of earthing is to avoid or minimize the danger of electrocution, fire due to earth leakage of current through undesired path and to ensure that the potential of a current carrying conductor does not rise with respect to the earth than its designed insulation.

Points to be Earthed: According to IE rules and IEE (Institute of Electrical Engineers) regulations, following points needs earthing.

- Earth pin of 3-pin lighting plug sockets and 4-pin power plug should be efficiently and permanently earthed.

- All metal casing or metallic coverings containing or protecting any electric supply line or apparatus such as GI pipes and conduits enclosing VIR or PVC cables, iron clad switches, iron clad distribution fuse boards etc should be earthed (connected to earth).

- The frame of every generator, stationary motors and metallic parts of all transformers used for controlling energy should be earthed by two separate and yet distinct connections with the earth.

- Every electrical panel or distribution box body should be earthed.

- Stay wires that are for overhead lines should be connected to earth by connecting at least one strand to the earth wires.

- Every server rack body should be earthed.

Earth Resistance

As per to IEEE rules, resistance between consumer earth terminal and earth Continuity conductor (at the end) should not be more than 1Ω. In case if you find the earth resistance more than 1Ω then you should immediate take action and maintain the earth resistance by undertaking maintenance on earth pit. The action can be adding salt water in earth pit.

Size of Earthing Wire should not be less than the half of the cross section area of thickest wire used in electrical wire installation. There should be minimum joints in earthing lead as well as lower in size and straight in the direction. Generally, copper wire can be used as earthing lead but, copper strip is also used for high installation and it can handle the high fault current because of wider area than the copper wire. A hard drawn bare copper wire is also used as an earthing lead. In this method, all earth conductors connected to a common (one or more) connecting points and then, earthing lead is used to connect earth electrode (earth plat) to the connecting point.

Earth Plate or Earth Pipe Size for small Installation In small installation you can use metallic rod (preferably Galvanised Iron) (diameter = 25mm (1inch) and length = 2m (6ft) instead of earth plate for earthing system. The metallic pipe should be 2 meter below from the surface of ground. To maintain the moister condition, put 25mm (1inch) coal and lime mixture around the earth plate. For effectiveness and convenience, you may use the copper rods 12.5mm (0.5 inch) to 25mm (1 inch) diameter and 4m (12ft) length.

Types of Earthing

Plate Earthing: In plate earthing system, a plate made up of either copper or GI is used.

a) The dimensions required for copper plate is: 60cm x 60cm x 3.18mm (i.e. 2ft x 2ft x 1/8 in)

b) And the dimensions required for galvanized iron (GI) of dimensions 60cm x 60cm x 6.35 mm (2ft x 2ft x ¼ in)

And is buried vertical in the earth (earth pit) which should not be less than 3m (10ft) from the ground level.

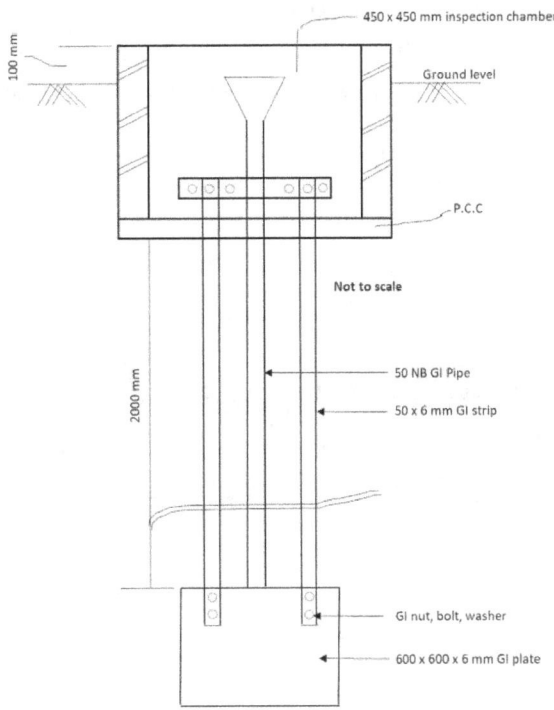

Pipe Earthing: In this kind of earthing a galvanized steel perforated pipe of approved length and diameter is placed vertically in a wet soil. It is the most common and cheaper system of earthing.

The size of pipe to use depends on the magnitude of current and the type of soil. The dimension of the pipe is usually 40mm (1.5in) in diameter and 2.75m (9ft) in length for ordinary soil or greater for dry and rocky soil. The moisture of the soil will determine the length of the pipe to be buried but usually it should be 4.75m (15.5ft).

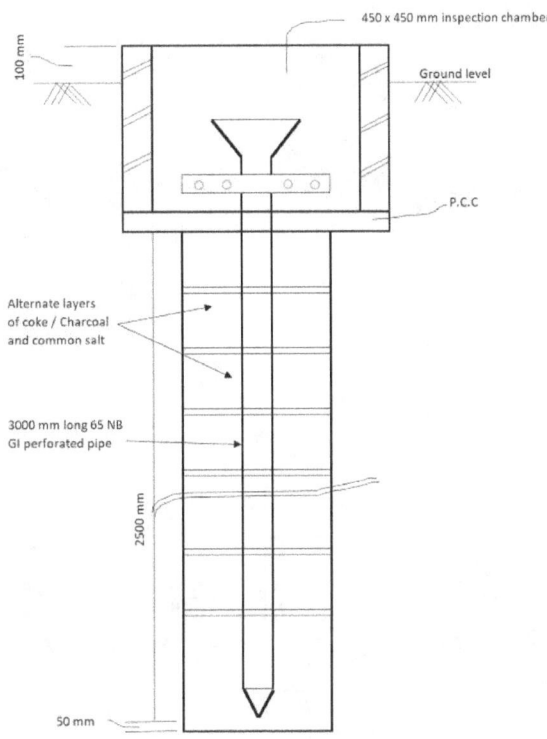

Rod Earthing: This kind of earthing is similar to pipe earthing. A copper rod of 12.5mm (1/2 inch) diameter or 16mm (0.6in) diameter or hollow section 25mm (1inch) of GI pipe of length above 2.5m (8.2 ft) are buried upright in the earth. The length of embedded electrodes in the soil reduces earth resistance to a desired value.

You may contact for online trainings & Corporate trainings at

ramesh@facilitymanagementinstitute.com

Ifme.facilitytraining@gmail.com

www.ingramcontent.com/pod-product-compliance
Lightning Source LLC
Chambersburg PA
CBHW081341180526
45171CB00006B/579